Studies in Systems, Decision and Control

Volume 209

The series "Studies in Systems, Decision and Control" (SSDC) covers both new developments and advances, as well as the state of the art, in the various areas of broadly perceived systems, decision making and control–quickly, up to date and with a high quality. The intent is to cover the theory, applications, and perspectives on the state of the art and future developments relevant to systems, decision making, control, complex processes and related areas, as embedded in the fields of engineering, computer science, physics, economics, social and life sciences, as well as the paradigms and methodologies behind them. The series contains monographs, textbooks, lecture notes and edited volumes in systems, decision making and control spanning the areas of Cyber-Physical Systems, Autonomous Systems, Sensor Networks, Control Systems, Energy Systems, Automotive Systems, Biological Systems, Vehicular Networking and Connected Vehicles, Aerospace Systems, Automation, Manufacturing, Smart Grids, Nonlinear Systems, Power Systems, Robotics, Social Systems, Economic Systems and other. Of particular value to both the contributors and the readership are the short publication timeframe and the world-wide distribution and exposure which enable both a wide and rapid dissemination of research output.

** Indexing: The books of this series are submitted to ISI, SCOPUS, DBLP, Ulrichs, MathSciNet, Current Mathematical Publications, Mathematical Reviews, Zentralblatt Math: MetaPress and Springerlink.

More information about this series at http://www.springer.com/series/13304

Mauricio A. Sanchez · Leocundo Aguilar ·
Manuel Castañón-Puga ·
Antonio Rodríguez
Editors

Applied Decision-Making

Applications in Computer Sciences and Engineering

Springer

Editors
Mauricio A. Sanchez
Universidad Autónoma de Baja California
Tijuana, Baja California, Mexico

Leocundo Aguilar
Universidad Autónoma de Baja California
Tijuana, Baja California, Mexico

Manuel Castañón-Puga
Universidad Autónoma de Baja California
Tijuana, Baja California, Mexico

Antonio Rodríguez
Universidad Autónoma de Baja California
Tijuana, Baja California, Mexico

ISSN 2198-4182 ISSN 2198-4190 (electronic)
Studies in Systems, Decision and Control
ISBN 978-3-030-17984-7 ISBN 978-3-030-17985-4 (eBook)
https://doi.org/10.1007/978-3-030-17985-4

This Springer imprint is published by the registered company Springer Nature Switzerland AG
The registered company address is: Gewerbestrasse 11, 6330 Cham, Switzerland

Preface

Within computer science there are far too many distinct fields of study that it would become impossible to enlist them all, yet one of the current trends which has been increasing in interest is the area of intelligent decision-making using computational tools. This book focuses on this distinct field, and more specifically, in applied decision-making. Where in this day and age, humanity has been giving more and more control over to computer systems in order to carry out a plethora of activities, releasing the human as a computational source and ceding control over to computers, more precisely, intelligent systems.

Even though this implies great challenges, many researchers have been proposing all sorts of solutions that can shed light on how to advance the field. It is clear that it is impossible to find solutions to every context in which a decision is taken by humans, but it is clear that sooner or later computer scientist will have enough proposals that will help figure out how general decision-making mechanisms can be implemented as a computational system.

The objective of this book is to portray a sample of advances in both applications and conceptualization of trends in applied decision-making. All included material is intended as a reference for students in pre-graduate level, wanting to know the state of the art as well as future directions this field might take for future professionals; as for graduate level, it is intended for students needing up to date reference materials as well as ideas for their research; and for professionals, who might want to be kept up to date in what current intelligent systems are capable of in order to either implement in their organizations or have a future view of possible turns their fields might take.

This book covers diverse accounts of applied decision-making, for example, security, optimization, complex systems, modeling and simulation, ubiquitous computing, among other specific trends, such that the contents within the covers of this book are sure to keep anyone interested, and we hope that readers enjoy its chapters as much as we enjoyed compiling and curing its entirety.

Tijuana, Mexico
February 2019

Mauricio A. Sanchez
Leocundo Aguilar
Manuel Castañón-Puga
Antonio Rodríguez

Contents

Evaluation of the Succession Measures of the Simultaneous Perturbation Stochastic Approximation Algorithm for the Optimization of the Process Capability Index

Juan Carlos Castillo García, Jesús Everardo Olguín Tiznado, Everardo Inzunza González, Claudia Camargo Wilson, Juan Andrés López Barreras and Enrique Efren García Guerrero

Abstract Simultaneous Perturbation Stochastic Approximation (SPSA) algorithms are alternative methods for optimizing systems where the relationship between the dependent variables and independent variables of a process is unknown. The objective of this research is to determine the optimum succession measure of SPSA that maximizes the Process Capability Index (PCI) through second order regression models by means of experimental simulation. The results show that three out of the ten combinations of the succession measures evaluated in SPSA yield optimum values that maximize the PCI according to the Six Sigma Methodology (DMAIC—Define, Measure, Analyze, Improve, and Control), this because the values have behaviors classified as world-class, this is, processes that generate less than 3.4 defects per million opportunities, which improves customer satisfaction and reduces cycle time and defects.

J. C. C. García · J. E. O. Tiznado (✉) · E. I. González · C. C. Wilson · E. E. G. Guerrero
Faculty of Engineering, Architecture, and Design, Autonomous University of Baja California, Carretera Transpeninsular Ensenada - Tijuana No. 3917 Colonia Playitas, 22860 Ensenada, Baja California, Mexico
e-mail: jeol79@uabc.edu.mx

J. C. C. García
e-mail: carlos.castillo@uabc.edu.mx

E. I. González
e-mail: einzunza@uabc.edu.mx

C. C. Wilson
e-mail: ccamargo@uabc.edu.mx

E. E. G. Guerrero
e-mail: eegarcia@uabc.edu.mx

J. A. L. Barreras
Faculty of Chemical Sciences and Engineering, Autonomous University of Baja California, Calzada Universidad #14418, Parque Industrial Internacional Tijuana, 22390 Tijuana, Baja California, Mexico
e-mail: jalopez@uabc.edu.mx

M. A. Sanchez et al. (eds.), *Applied Decision-Making*, Studies in Systems, Decision and Control 209, https://doi.org/10.1007/978-3-030-17985-4_1

Keywords SPSA · PCI · Optimization · Six Sigma · DMAIC

1 Introduction

The Simultaneous Perturbation Stochastic Approximation (SPSA) Algorithm proposed by Spall [57] is a multivariable stochastic approximation procedure of the Kiefer and Wolfowitz type [15, 32, 46, 47]. They do not require a deep knowledge of the system to be optimized, that is, they are applicable in situations in which the functional relationship between the response variable to be optimized represented by $\hat{y}$ and its controllable variables represented by X_k, García [23].

This algorithm is the most efficient of the Kiefer and Wolfowitz types, both theoretically and practically, given that it presents a higher convergence rate and requires only two observations in each iteration to estimate the gradient [18, 28, 41, 44]. The latter is of great interest since the economic cost and the simplicity of the experimental work depend directly on it [15, 22, 46]. However, the SPSA Algorithm does not propose what to do in the case of its application in practice when in some iteration of this algorithm a new solution with one or more controllable X_k factors outside the feasible region or domain thereof is generated. Sadegh and Spall [52], Ito and Dhaene [27], Haghighat Sefat et al. [25] and Wang and Shi [65] solve this problem by developing the SPSA Algorithm with projection.

To guarantee the convergence of stochastic approximation procedures of the Kiefer-Wolfowitz type, it is necessary to define two decreasing sequences of non-negative real numbers a_k and c_k, of which the performance of this type of stochastic approach procedures depends on the way in which these successions are defined, for which, Sadegh and Spall [52] propose a criterion to guarantee their convergence.

To see numerical examples showing the dependence of stochastic approximation procedures of the Kiefer-Wolfowitz type of this with different combinations of these sequences, see the following works: Sadegh and Spall [52], L'Ecuyer et al. [34], Andradóttir [5], Broadie et al. [12] (2009) and Śliwiński and Wachel [56].

On the other hand, the analysis of the actual Process Capability Index ($\widehat{C_{pk}}$) consists of determining the operating conditions of the processes in which the average value of the response variable is closest to the objective value and that its variance has the minimum value [6, 20, 31, 35, 40, 45, 67, 11]. As a measure of performance, these indexes provide an effective way to measure whether the process meets the specification limits [1, 3, 6, 7, 17, 26, 29, 36, 40, 48, 50, 63, 68, 20, 30, 33, 62, 54, 37]. In addition, it has the advantages that it is easy to estimate, and the information it provides is concrete and of great significance, which is why it has become the most popular and used capacity index in real-world applications as show in: Kane [30], Ramakrishnan et al. [48], Rodriguez [50], Tang and Wang [63], Ahmad et al. [3], Bai et al. [7], Méndez-González et al. [40] and de Felipe and Benedito [17].

García [23] presents an algorithm for the experimental improvement of the $(\widehat{C_{pk}})$ algorithm that uses the (SPSA) Algorithm with Sadegh projection [52], and the latter consists of the Spall algorithm [57] with projections on the domain of independent variables when they come out of that domain in the implementation of the (SPSA) Algorithm. In these works the convergence with probability one of the successions generated by these algorithms towards the optimal solution is demonstrated. However, in these same works, the behavior in the first iterations is not studied with different combinations of the sequences a_k and c_k, in terms of the number of iterations that are required to achieve a specific improvement in the value of the $(\widehat{C_{pk}})$, with the consequent savings in time and costs in the implementation of this type of algorithms in the improvement of the value of this actual PCI.

Therefore, this chapter presents an evaluation of the efficiency of the (SPSA) Algorithm for different combinations of sequences a_k and c_k, required in its implementation, with the objective of determining with which of these combinations it is possible to maximize the actual PCI $(\widehat{C_{pk}})$ with the smallest number of iterations, and with this, achieve in an economical way to increase the production within the specifications and reduce the generation of defective parts per million to the minimum.

The rest of this work is organized in four sections: Sect. 2 presents the materials and methods used in the adaptation of the (SPSA) Algorithm for the optimization of $(\widehat{C_{pk}})$, Sect. 3 presents the analysis of results based in the simulation of four numerical examples to scale, and finally, Sect. 4 presents the conclusions and directions for future research.

2 Materials and Methods

In this section, we present the materials used in the development of this article, which is based on an optimization work by experimental simulation, developed with a laptop computer with AMD E-450 processor with 1.65 GHz and 4 GB RAM. In addition, the software to carry out the statistical analysis of the data were Minitab 17® to perform the calculation of the second order models, and MATLAB R2014® to carry out the simulation of the Simultaneous Perturbation Stochastic Approximation Algorithm for the improvement of the actual PCI $(\widehat{C_{pk}})$.

2.1 SPSA Algorithm and Optimization of the PCI $(\widehat{C_{pk}})$

Next, the method used to obtain the necessary data for the experimental simulation design and analysis of this chapter will be described.

Step 0. Obtain the second order dual model for the mean (μ) and standard deviation (σ) by using a classic 3^k experiment design or a Central Composite Design (CCD)

Montgomery [42]. In this chapter we used the second order regression models for the mean (μ) and standard deviation (σ) published by Olguín [43], see page 103, model for industrial process 1 and Gedi et al. [24], see page 5, CCD with five central points for the models of the Yield, EPA and DHA evaluation processes). In this chapter, four numerical examples selected from the literature are presented to carry out the experimental simulation analysis as shown below: *First*, the dual second-order regression model reported by Olguín [43] is presented, which measures the percentage of final moisture in the dehydrated garlic process with its target value of five percent of final moisture.

$$\widehat{y_{\mu 1}} = 5.915 + 0.137X_1 + 1.893X_2 - 0.002X_1^2 - 0.031X_1X_2 - 0.04X_2^2,$$
$$\widehat{y_{\sigma 1}} = -18.585 + 0.749X_1 - 2.362X_2 - 0.002X_1^2 - 0.115X_1X_2 + 1.065X_2^2.$$

Second, dual models of second-order regression are presented, calculated based on the results reported by Gedi et al. [24], which measure the performance of the sardine oil extraction process (YIELD = $\widehat{y_{\mu 2}}$), and the relationships of EicosaPentaenoic Acid (EPA = $\widehat{y_{\mu 3}}$), and DocosaHexaenoic Acid (DHA = $\widehat{y_{\mu 4}}$). Therefore, the Yield measures the percentage of the total extraction yield and the ratio of the EicosaPentaenoic and DocosaHexaenoic acids. The models are presented below for the mean and for the standard deviation in each of the evaluated responses:

$$\widehat{y_{\mu 2}} = -34.26 + 0.1504X_1 + 0.6631X_2 - 0.000184X_1^2 - 0.000572X_1X_2 - 0.004413X_2^2,$$
$$\widehat{y_{\sigma 2}} = 0.227 + 0.00127X_1 - 0.0125X_2 + 0.00001X_1^2 - 0.000141X_1X_2 + 0.000555X_2^2.$$

The previous expressions were adjusted with correlation coefficients $r^2 = 97.76\%$ for the mean model and 95.45% for the deviation model:

$$\widehat{y_{\mu 3}} = 7.972 + 0.0033X_1 + 0.06349X_2 - 0.000009X_1^2 - 0.000187X_1X_2 - 0.000087X_2^2,$$
$$\widehat{y_{\sigma 3}} = -0.153 + 0.00105X_1 + 0.00009X_2 - 0.000001X_1^2 - 0.000007X_1X_2 + 0.000029X_2^2.$$

In the case of the EPA, the correlation coefficients were $r^2 = 98.77\%$ for the mean model and 42.35% for the deviation model:

$$\widehat{y_{\mu 4}} = 12.796 + 0.03005X_1 + 0.1377X_2 - 0.000029X_1^2 - 0.000137X_1X_2 - 0.001879X_2^2,$$
$$\widehat{y_{\sigma 4}} = -0.07 - 0.00103X_1 + 0.0106X_2 + 0.000005X_1^2 - 0.000026X_1X_2 - 0.000023X_2^2.$$

The correlation coefficients for DHA considered were: $r^2 = 99.66\%$ for the mean model and 22.2% for the standard deviation model.

Step 1. Initialization and selection coefficient. Select the index counter $k = 1$ and take a presumed value of the initial gradient vector and the coefficients of non-negativity a, c, A, α and γ. Choose the measures of a_k and c_k, values that are critical for the performance of the SPSA Algorithm. Andradóttir [5], Spall [58] and Altaf et al. [4] establish that the value that is typically assumed for $a_k = a/k$ and for $c_k = c/(k+1)^{\gamma}$ when the gradient vector is equal to the arithmetic mean of m estimates.

However, Spall [59] establishes that these measures of succession a_k and c_k are proposed parameters, developed on the basis of many case studies, but may vary, since there are not a couple of these measures of succession that are robust for any type of application.

Therefore, in this research, we evaluated at least five measures of succession a_k and two of c_k to evaluate the efficiency of the (SPSA) Algorithm, which were found in the literature that we will list below: $a_{k1} = a/(A + k)^{\alpha}$ Spall [60], Spall [59], Sidorov et al. [55], Reardon et al. [49], McClary et al. [39], Cipriani et al. [16] and Spall [61]; $a_{k2} = a/(A + k + 1)^{\alpha}$ Spall [58], Venkatesh et al. [64], Altaf et al. [4], Wang [66], Abdulsadda and Iqbal [2], Seyedpoor et al. [53], Altaf et al. [4] and Ito and Dhaene [27], Ros-Roca et al. [51], Wang and Shi [65]; $a_{k3} = a/k$ Baltcheva et al. [8] and Bartkute and Sakalauskas [10]; $a_{k4} = a/k^{2/3}$ Garcia [23]; $a_{k5} = a/k^{\alpha}$ Sadegh and Spall [52] and Bangerth et al. [9]; and for c_k are $c_{k1} = c/k^{\gamma}$ Spall [59], Baltcheva et al. [8], Bartkute and Sakalauskas [10], Bangerth et al. [9], Reardon et al. [49] and McClary et al. [39]; $c_{k2} = c/(k + 1)^{\gamma}$ Spall [58], Sidorov et al. [55], Venkatesh et al. [64], Cao [14], Altaf et al. [4], Seyedpoor et al. [53], Abdulsadda and Iqbal [2], Spall [61], Ding et al. [19], Altaf et al. [4] and Ito and Dhaene [27], Ros-Roca et al. [51], Wang and Shi [65].

Where: a is the initial value of the step size of the gradient factor, c is the initial value of the step size in the approximation of the gradient factor, A is the Stability factor, k is the iteration number that is evaluates, α is the rate of reduction of the step size of the gradient factor, γ is the rate of reduction of the size of the step in the approximation of the gradient factor [21].

Values practically effective and theoretically valid for α and γ are 0.602 and 0.101 respectively (the optimal asymptotic values of 1.0 and 1/6 can also be used) [4, 43, 58, 59, 66]. A useful guide when choosing A is to do it as if it were much smaller than the maximum number of iterations allowed or expected, that is why $A = 100$, $\alpha = 0.16$ and $c = 1$ was selected.

Step 2. Generate the Simultaneous Perturbation Vector. It is generated by the Montecarlo method, a p-dimensional random perturbation vector Δ_k, where each of the p components of Δ_k is generated independently of a probability distribution with zero mean. One option for each of the components of Δ_k is to use a Bernoulli distribution ± 1 with a probability of 0.5 for each result ± 1. Note that uniform and normal random variables are not allowed for the elements of Δ_k by the regular conditions of the (SPSA) Algorithm according to Brooks [13], Maryak and Chin [38]. Therefore, the values of the perturbation vector for the experimental simulation design and analysis of this article are the following: for the model reported by Olguín [43] $\Delta_k^+ = 3\,^{\circ}\mathrm{C}$ and $\Delta_k^- = -0.3\,\mathrm{h}$, and for he models calculated with the results reported by Gedi et al. [24] the $\Delta_k^+ = 141.4$ Bars and $\Delta_k^- = -21.5\,^{\circ}\mathrm{C}$.

Step 3. Evaluate the function to maximize the PCI ($\widehat{C_{pk}}$).

3.a. Select initial values for $X_k : X_1 = 65\,°\text{C}$ and $X_2 = 4\,\text{h} : 30\,\text{min}$ [43; page 103 industrial process 1], and for the analysis of the models [24] $X_1 = 300$ Bars and $X_2 = 55\,°\text{C}$.

3.b. After selecting X_k, we replace the values corresponding to the independent variables (X_1, X_2) of the second order regression models $\widehat{y_{\mu 1}}$ and $\widehat{y_{\sigma 1}}$, $\widehat{y_{\mu 2}}$ and $\widehat{y_{\sigma 2}}$, $\widehat{y_{\mu 3}}$ and $\widehat{y_{\sigma 3}}$, and $\widehat{y_{\mu 4}}$ and $\widehat{y_{\mu 4}}$ reported in **step 0**.

3.c. Obtained the values of $\widehat{y_{\mu 1}} = 5.011$, $\widehat{y_{\sigma 1}} = -1.050$, $\widehat{y_{\mu 2}} = 7.983$, $\widehat{y_{\sigma 2}} = 0.173$, $\widehat{y_{\mu 3}} = 8.295$, $\widehat{y_{\sigma 3}} = 0.049$, $\widehat{y_{\mu 4}} = 18.83$ and $\widehat{y_{\sigma 4}} = 0.1554$, calculate the estimated value of the $\widehat{C_{pk}}(X_k)$. Where we replace the values of $\widehat{y_{\mu 1}}$ and $\widehat{y_{\sigma 1}}$, $\widehat{y_{\mu 2}}$ and $\widehat{y_{\sigma 2}}$, $\widehat{y_{\mu 3}}$ and $\widehat{y_{\sigma 3}}$, and $\widehat{y_{\mu 4}}$ and $\widehat{y_{\sigma 4}}$ in Eq. (1) to calculate the PCI $\widehat{C_{pk}}(X_k)$ as follows:

$$\begin{aligned}
\widehat{C_{pk}}(X_k) &= \min\left\{\frac{USL - \widehat{y_\mu}}{3 * \widehat{y_\sigma}}, \frac{\widehat{y_\mu} - 4}{3 * \widehat{y_\sigma}}\right\},\\
\widehat{C_{pk1}}(X_{k1}) &= \min\left\{\frac{6 - 5.011}{3 * (-1.05)}, \frac{5.011 - 4}{(3 * -1.05)}\right\},\\
\widehat{C_{pk2}}(X_{k2}) &= \min\left\{\frac{10 - 7.983}{3 * 0.173}, \frac{7.983 - 5}{3 * 0.173}\right\},\\
\widehat{C_{pk3}}(X_{k3}) &= \min\left\{\frac{10 - 8.295}{3 * 0.049}, \frac{8.295 - 4}{3 * 0.049}\right\},\\
\widehat{C_{pk4}}(X_{k4}) &= \min\left\{\frac{21 - 18.83}{3 * 0.4346}, \frac{18.83 - 16}{3 * 0.4346}\right\}.
\end{aligned} \tag{1}$$

where $\widehat{y_\mu}$ represents the observed mean for the response variable; USL and LSL, represent the Upper and Lower Specification Limits respectively and $\widehat{y_\sigma}$ represents the variation observed for the response variable. Then, the values for the numerical analysis 1 are $\text{USL}_1 = 6$ and $\text{LSL}_1 = 4\%$ humidity respectively; for the numerical analyzes 2, 3 and 4 are $\text{USL}_2 = 10$ and $\text{LSL}_2 = 5\%$, $\text{USL}_3 = 10$ and $\text{LSL}_3 = 4\%$, $\text{USL}_4 = 21$ and $\text{LSL}_4 = 16\%$. Therefore, the value of $\widehat{C_{pk1}}(X_{k1}) = -0.321$, $\widehat{C_{pk2}}(X_{k2}) = 3.888$, $\widehat{C_{pk3}}(X_{k3}) = 11.555$ and $\widehat{C_{pk4}}(X_{k4}) = 4.653$..

3.d. Calculate the values of the controllable factors X_k^+, which is based on the initial value of X_k, the measure of sequence c_k and the perturbation vector $\Delta_k^{\pm}$; these are used to obtain the values of $X_k^+ = X_k + c_k(\Delta_k^{\pm})$.

The new values of the controllable factors X_k^+ are $X_1^+ = 67.8\,°\text{C}$, $X_2^+ = 4\,\text{h}\ 13\,\text{min}$, for model 1; $X_1^+ = 432$ Bars, $X_2^+ = 35\,°\text{C}$, for model 2, model 3 and model 4 where these values are substituted in the dual models of second order regression for the following mean and standard deviation:

$$\begin{aligned}
\widehat{y_{\mu 1}^+} &= 5.915 + 0.137X_1^+ + 1.893X_2^+ - 0.002X_1^{2+} - 0.031X_1^+X_2^+ - 0.04X_2^{2+},\\
\widehat{y_{\sigma 1}^+} &= -18.585 + 0.749X_1^+ - 2.362X_2^+ - 0.002X_1^{2+} - 0.115X_1^+X_2^+ + 1.065X_2^{2+}.
\end{aligned}$$

$$\widehat{y_{\mu 2}^{+}} = -34.26 + 0.1504X_1^{+} + 0.6631X_2^{+} - 0.000184X_1^{2+} - 0.000572X_1^{+}X_2^{+} - 0.004413X_2^{2+},$$

$$\widehat{y_{\sigma 2}^{+}} = 0.227 + 0.00127X_1^{+} - 0.0125X_2^{+} + 0.00001X_1^{2+} - 0.000141X_1^{+}X_2^{+} + 0.000555X_2^{2+}.$$

$$\widehat{y_{\mu 3}^{+}} = 7.972 + 0.0033X_1^{+} + 0.06349X_2^{+} - 0.000009X_1^{2+} - 0.000187X_1^{+}X_2^{+} - 0.000087X_2^{2+},$$

$$\widehat{y_{\sigma 3}^{+}} = -0.153 + 0.00105X_1^{+} + 0.00009X_2^{+} - 0.000001X_1^{2+} - 0.000007X_1^{+}X_2^{+} + 0.000029X_2^{2+}.$$

$$\widehat{y_{\mu 4}^{+}} = 12.796 + 0.03005X_1^{+} + 0.1377X_2^{+} - 0.000029X_1^{2+} - 0.000137X_1^{+}X_2^{+} - 0.001879X_2^{2+},$$

$$\widehat{y_{\sigma 4}^{+}} = -0.07 - 0.00103X_1^{+} + 0.0106X_2^{+} + 0.000005X_1^{2+} - 0.000026X_1^{+}X_2^{+} - 0.000023X_2^{2+}.$$

Obtaining the results of the values of $\widehat{y_{\mu 1}^{+}} = 4.417$, $\widehat{y_{\sigma 1}^{+}} = -0.902$, $\widehat{y_{\mu 2}^{+}} = 5.535$, $\widehat{y_{\sigma 2}^{+}} = 0.237$, $\widehat{y_{\mu 3}^{+}} = 7.012$, $\widehat{y_{\sigma 3}^{+}} = 0.449$, $\widehat{y_{\mu 4}^{+}} = 20.814$ and $\widehat{y_{\sigma 4}^{+}} = 0.367$.

3.e. Calculate the estimated value of the $\widehat{C_{pk}^{+}(X_k^{+})}$. Replace the values of $\widehat{y_{\mu 1}^{+}} = 4.417$, $\widehat{y_{\sigma 1}^{+}} = -0.902$, $\widehat{y_{\mu 2}^{+}} = 5.535$, $\widehat{y_{\sigma 2}^{+}} = 0.237$, $\widehat{y_{\mu 3}^{+}} = 7.012$, $\widehat{y_{\sigma 3}^{+}} = 0.449$, $\widehat{y_{\mu 4}^{+}} = 20.814$ and $\widehat{y_{\sigma 4}^{+}} = 0.367$ in Eq. (2) to calculate the PCI $\widehat{C_{pk}^{+}(X_k^{+})}$ as follows:

$$\widehat{C_{pk1}^{+}(X_{k1}^{+})} = \min\left\{\frac{USL - \widehat{y_{\mu 1}^{+}}}{3\widehat{y_{\sigma 1}^{+}}}; \frac{\widehat{y_{\mu 1}^{+}} - LSL}{3\widehat{y_{\sigma 1}^{+}}}\right\}$$

$$\widehat{C_{pk1}^{+}(X_{k1}^{+})} = \min\left\{\frac{6 - 4.417}{3 * (-0.902)}, \frac{4.417 - 4}{(3 * -0.902)}\right\}$$

$$\widehat{C_{pk2}^{+}(X_{k2}^{+})} = \min\left\{\frac{10 - 5.535}{3 * 0.237}, \frac{5.535 - 5}{3 * 0.237}\right\}$$

$$\widehat{C_{pk3}^{+}(X_{k3}^{+})} = \min\left\{\frac{10 - 7.012}{3 * 0.449}, \frac{7.012 - 4}{3 * 0.449}\right\}$$

$$\widehat{C_{pk4}^{+}(X_{k4}^{+})} = \min\left\{\frac{21 - 20.814}{3 * 0.367}, \frac{20.814 - 16}{3 * 0.367}\right\} \tag{2}$$

Therefore, considering the values of the USL and LSL of Eq. 1, the value of the $\widehat{C_{pk1}^{+}(X_{k1}^{+})} = -0.585$, $\widehat{C_{pk2}^{+}(X_{k2}^{+})} = 0.237$, $\widehat{C_{pk3}^{+}(X_{k3}^{+})} = 2.218$ and $\widehat{C_{pk4}^{+}(X_{k4}^{+})} = 0.168$.

3.f. Calculate the values of the controllable factors X_k^{-}, which is based on the initial value of X_k, the measurement of succession c_k and the perturbation vector $\Delta_k^{\pm}$; these are used to obtain the values of $X_k^{-} = X_k - c_k(\Delta_k^{\pm})$. The new values of the controllable factors X_k^{-} are $X_1^{-} = 62.2\,°C$, $X_2^{-} = 4\,h\ 47\,min$, where these values are substituted in the dual models of second order regression for the following mean and standard deviation:

$$\widehat{y_{\mu 1}^{-}} = 5.915 + 0.137X_1^{-} + 1.893X_2^{-} - 0.002X_1^{2-} - 0.031X_1^{-}X_2^{-} - 0.04X_2^{2-}$$

$$\widehat{y_{\sigma 1}^{-}} = -18.585 + 0.749X_1^{-} - 2.362X_2^{-} - 0.002X_1^{2+} - 0.115X_1^{-}X_2^{-} + 1.065X_2^{2-}$$

$$\widehat{y_{\mu 2}^{-}} = -34.26 + 0.1504X_1^{-} + 0.6631X_2^{-} - 0.000184X_1^{2-} - 0.000572X_1^{-}X_2^{-} - 0.004413X_2^{2-}$$

$$\widehat{y_{\sigma 2}^{-}} = 0.227 + 0.00127X_1^- - 0.0125X_2^- + 0.00001X_1^{2-} - 0.000141X_1^- X_2^- + 0.000555X_2^{2-}$$

$$\widehat{y_{\mu 3}^{-}} = 7.972 + 0.0033X_1^- + 0.06349X_2^- - 0.000009X_1^{2-} - 0.000187X_1^- X_2^- - 0.000087X_2^{2-}$$

$$\widehat{y_{\sigma 3}^{-}} = -0.153 + 0.00105X_1^- + 0.00009X_2^- - 0.000001X_1^{2-} - 0.000007X_1^- X_2^- + 0.000029X_2^{2-}$$

$$\widehat{y_{\mu 4}^{-}} = 12.796 + 0.03005X_1^- + 0.1377X_2^- - 0.000029X_1^{2-} - 0.000137X_1^- X_2^- - 0.001879X_2^{2-}$$

$$\widehat{y_{\sigma 4}^{-}} = -0.07 - 0.00103X_1^- + 0.0106X_2^- + 0.000005X_1^{2-} - 0.000026X_1^- X_2^- - 0.000023X_2^{2-}$$

3.g. Calculate the estimated value of the $\widehat{C_{pk}^{-}}(X_k^-)$. Replace the values of $\widehat{y_{\mu 1}^{-}} = 5.616$, $\widehat{y_{\sigma 1}^{-}} = -0.883$, $\widehat{y_{\mu 2}^{-}} = 3.532$, $\widehat{y_{\sigma 2}^{-}} = 1.123$, $\widehat{y_{\mu 3}^{-}} = 10.183$, $\widehat{y_{\sigma 3}^{-}} = 0.28$, $\widehat{y_{\mu 4}^{-}} = 15.054$, and $\widehat{y_{\sigma 4}^{-}} = 0.235$ in Eq. (3) to calculate the process capability index $\widehat{C_{pk}^{-}}(X_k^-)$ as follows:

$$\widehat{C_{pk}^{-}}(X_k^-) = \min\left\{\frac{USL - \widehat{y_{\mu 1}}}{3\widehat{y_{\sigma 1}}}; \frac{\widehat{y_{\mu 1}} - LSL}{3\widehat{y_{\sigma 1}}}\right\}$$
$$\widehat{C_{pk1}^{-}}(X_{k1}^-) = \min\left\{\frac{6 - 5.616}{3*(-0.883)}, \frac{5.616 - 4}{(3*-883)}\right\}$$
$$\widehat{C_{pk2}^{-}}(X_{k2}^-) = \min\left\{\frac{10 - 3.532}{3*1.123}, \frac{3.532 - 5}{3*1.123}\right\}$$
$$\widehat{C_{pk3}^{-}}(X_{k3}^-) = \min\left\{\frac{10 - 10.183}{3*0.280}, \frac{10.183 - 4}{3*0.280}\right\}$$
$$\widehat{C_{pk4}^{-}}(X_{k4}^-) = \min\left\{\frac{21 - 15.054}{3*0.235}, \frac{15.054 - 16}{3*0.235}\right\} \tag{3}$$

Therefore, considering the values of the LES and LEI of Eq. 1, the value of the $\widehat{C_{pk1}^{-}}(X_{k1}^-) = -0.610$, $\widehat{C_{pk2}^{-}}(X_{k2}^-) = -0.433$, $\widehat{C_{pk3}^{-}}(X_{k3}^-) = -0.218$ and $\widehat{C_{pk4}^{-}}(X_{k4}^-) = -1.337$.

Step 4. Gradient approximation for $\emptyset_k^+$ y $\emptyset_k^-$. Generate the simultaneous perturbation approximation of the unknown gradient as follows:

$$\emptyset_k^+ = \frac{\widehat{C_{pk}^{+}}(X_k^+) - \widehat{C_{pk}^{-}}(X_k^-)}{2c_k\Delta_k^+} \tag{4}$$

$$\emptyset_k^- = \frac{\widehat{C_{pk}^{+}}(X_k^+) - \widehat{C_{pk}^{-}}(X_k^-)}{2c_k\Delta_k^-} \tag{5}$$

Resulting in a value of $\emptyset_k^+ = 0.0041$, a value of $\emptyset_k^- = -0.0418$ for model 1; $\emptyset_k^+ = 0.0076$ and $\emptyset_k^- = -0.0499$ for model 2; $\emptyset_k^+ = -0.012$ and $\emptyset_k^- = 0.0791$ for model 3; $\emptyset_k^+ = 0.0218$ and $\emptyset_k^- = -0.1436$ for model 4; so we proceed to obtain the values of $\widehat{X}_k$.

Step 5. Update the value of the controllable factors $\widehat{X}_k$, estimating a new value for the iterative process of $\widehat{X}^{+}_{k+1}$ and $\widehat{X}^{-}_{k+1}$ this is obtained by using the standard Stochastic Algorithm formula as follows:

$$\widehat{X}^{+}_{k+1} = X_k - a_k \emptyset^{+}_{k} \quad (6)$$

$$\widehat{X}^{-}_{k+1} = X_k - a_k \emptyset^{-}_{k} \quad (7)$$

resulting in a value for $\widehat{X}^{+}_{k+1} = 64.99$ and a value of $\widehat{X}^{-}_{k+1} = 4.5009$, for model 1; $\widehat{X}^{+}_{k+1} = 299.9999$ and $\widehat{X}^{-}_{k+1} = 55.0004$, for model 2; $\widehat{X}^{+}_{k+1} = 300.000$ and $\widehat{X}^{-}_{k+1} = 55.000$, for model 3 and model 4; so we proceed to obtain the values of the iterative process considered for 100, 200, and 500 iterations.

Step 6. Iteration or Termination. Return to step 2 with $k + 1$ replacing k. Finish the algorithm when the value of the actual process capability index is obtained $\widehat{C_{pk}} \geq 1.33$ or the maximum number of iterations has been rejected.

3 Results

In this section, it is presented the results obtained from the experiments carried out with the succession measurements of the SPSA Algorithm in order to evaluate their efficiency with different combinations of a_k and c_k sequences required in their implementation, this, with the purpose to determine which of the combinations of the succession measurements manages to maximize the actual process capability index $\widehat{C_{pk}}$ in the least amount of iterations or possible experiments. Therefore, the combinations of experiments of the succession measurements within the SPSA Algorithm are presented in Table 1.

The numerical results obtained from the iterative processes generated by the SPSA Algorithm by means of simulation for the model reported by Olguín et al. [43, see page 103, model for industrial process 1] are presented in Tables 2, 3, and 4. Where the evaluated experiment, the combination of the succession measurements, the optimal values found for the controllable factors, the maximum value of the response variable ($\widehat{C_{pk}}$) is specified, and finally, the number of iterations where the maximum value of the response variable was obtained. Likewise, the results obtained from the iterative processes for the Yield, EPA and DHA models reported by Gedi et al. [24] are shown in Tables 5, 6, 7, 8, and 9

Next, in Table 2, we show the maximum values of the response variable ($\widehat{C_{pk}}$) in 100 iterations for the model reported by Olguín et al. [43, see page 103, model for industrial process 1].

As can be seen in Table 2, experiments 3, 8 and 9 fulfill the objectives sought in this research work, which are: obtain values of $\widehat{C_{pk}} \geq 1.33$ in the least number of

Table 1 Experiments on the succession measurements of the SPSA Algorithm

Experiments	a_k	c_k
1	$a_{k1} = \frac{a}{(A+k)^\alpha}$	$C_{k1} = \frac{c}{k^\gamma}$
2	$a_{k2} = \frac{a}{(A+k+1)^\alpha}$	
3	$a_{k3} = \frac{a}{k}$	
4	$a_{k4} = \frac{a}{k^{2/3}}$	
5	$a_{k5} = \frac{a}{k^\alpha}$	
6	$a_{k1} = \frac{a}{(A+k)^\alpha}$	$C_{k2} = \frac{c}{(1+k)^\gamma}$
7	$a_{k2} = \frac{a}{(A+k+1)^\alpha}$	
8	$a_{k3} = \frac{a}{k}$	
9	$a_{k4} = \frac{a}{k^{2/3}}$	
10	$a_{k5} = \frac{a}{k^\alpha}$	

Table 2 Results of the $\widehat{C_{pk}}$ evaluated by the SPSA Algorithm in the simulation process from 1 to 100 iterations in model reported by (Olguín et al. [43], page 103 industrial process 1)

Experiments	Combination of succession measurements in the SPSA Algorithm		Controllable factors		Response variable	Number of iterations
	a_k	c_k	X_1^*	X_2^*	$\widehat{C_{pk}}$max	Position
1	1	1	64.9976	4.5243	−0.3158	55
2	2	1	64.9976	4.5242	−0.3158	55
3	3	1	62.5239	5.4719	5.6935	9
4	4	1	62.5132	6.2598	0.1148	12
5	5	1	64.9977	4.5231	−0.3160	5
6	1	2	64.9976	4.5242	−0.3158	55
7	2	2	64.9976	4.5241	−0.3158	55
8	3	2	62.7556	5.4877	5.8698	24
9	4	2	64.8867	5.6333	2.9784	7
10	5	2	64.9973	4.5265	−0.3160	6

Table 3 Results of the $(\widehat{C_{pk}})$ evaluated by the SPSA Algorithm in the simulation process from 1 to 200 iterations in model reported by (Olguín et al. [43], page 103 industrial process 1)

Experiments	Combination of succession measurements in the SPSA Algorithm		Controllable factors		Response variable	Number of iterations
	a_k	c_k	X_1^*	X_2^*	$\widehat{C_{pk}}\max$	Position
1	1	1	64.9976	4.5243	−0.3158	55
2	2	1	64.9976	4.5242	−0.3158	55
3	3	1	62.5239	5.4719	5.6935	9
4	4	1	62.5132	6.2598	0.1148	12
5	5	1	63.1585	5.5048	10.7603	198
6	1	2	64.9976	4.5242	−0.3158	55
7	2	2	64.9976	4.5241	−0.3158	55
8	3	2	63.1325	5.4952	23.6241	169
9	4	2	64.8867	5.6333	2.9784	7
10	5	2	63.1384	5.5075	8.4123	176

Table 4 Results of $(\widehat{C_{pk}})$ evaluated by the SPSA Algorithm in the simulation process from 1 to 500 iterations in model reported by (Olguín et al. [43], page 103 industrial process 1)

Experiments	Combination of succession measurements in the SPSA Algorithm		Controllable factors		Response variable	Number of iterations
	a_k	c_k	X_1^*	X_2^*	$(\widehat{C_{pk}})$	Position
1	1	1	64.9976	4.5243	−0.3158	55
2	2	1	64.9976	4.5242	−0.3158	55
3	3	1	62.5239	5.4719	5.6935	9
4	4	1	62.5132	6.2598	0.1148	12
5	5	1	63.2969	5.5026	48.2164	449
6	1	2	64.9976	4.5242	−0.3158	55
7	2	2	64.9976	4.5241	−0.3158	55
8	3	2	63.2067	5.4955	78.3644	259
9	4	2	64.8867	5.6333	2.9784	7
10	5	2	63.2005	5.4969	41.2858	250

Table 5 Results of the $\widehat{C_{pk}}$ evaluated by the SPSA Algorithm in the simulation process from 1 to 100 iterations in the YIELD model reported by Gedi et al. [24]

Experiments	Combination of succession measurements in the SPSA Algorithm		Controllable factors		Response variable	Number of iterations
	a_k	c_k	X_1^*	X_2^*	$(\widehat{C_{pk}})$	Position
1	1	1	300	55	3,889	1
2	2	1	300	55	3,889	1
3	3	1	300	55	3,889	1
4	4	1	300	55	3,889	1
5	5	1	300	55	3,889	1
6	1	2	300	55	3,889	1
7	2	2	300	55	3,889	1
8	3	2	300	55	3,889	1
9	4	2	300	55	3,889	1
10	5	2	300	55	3,889	1

Table 6 Results of the $\widehat{C_{pk}}$ evaluated by the SPSA Algorithm in the simulation process from 1 to 100 iterations in the EPA model reported by Gedi et al. [24]

Experiments	Combination of succession measurements in the SPSA Algorithm		Controllable factors		Response variable	Number of iterations
	a_k	c_k	X_1^*	X_2^*	$(\widehat{C_{pk}})$	Position
1	1	1	388.593	41.666	20.705	100
2	2	1	388.593	41.665	20.705	100
3	3	1	388.458	42.554	20.891	100
4	4	1	388.209	44.190	21.169	100
5	5	1	388.549	41.954	20.768	100
6	1	2	388.504	41.680	20.707	100
7	2	2	388.504	41.679	20.707	100
8	3	2	388.366	42.590	20.897	100
9	4	2	388.115	44.237	21.173	100
10	5	2	388.460	41.972	20.771	100

Table 7 Results of the $\widehat{C_{pk}}$ evaluated by the SPSA Algorithm in the simulation process from 1 to 200 iterations in the EPA model reported by Gedi et al. [24]

Experiments	Combination of succession measurements in the SPSA Algorithm		Controllable factors		Response variable	Number of iterations
	a_k	c_k	X_1^*	X_2^*	$(\widehat{C_{pk}})$	Position
1	1	1	382.581	42.716	20.851	200
2	2	1	382.581	42.715	20.851	200
3	3	1	382.438	43.651	20.994	200
4	4	1	382.716	46.016	21.238	185
5	5	1	382.528	43.064	20.908	200
6	1	2	382.539	42.723	20.852	200
7	2	2	382.539	42.722	20.852	200
8	3	2	382.393	43.680	20.998	200
9	4	2	382.759	46.025	21.239	183
10	5	2	382.485	43.075	20.909	200

Table 8 Results of the $\widehat{C_{pk}}$ evaluated by the SPSA Algorithm in the simulation process from 1 to 500 iterations in the EPA model reported by Gedi et al. [24]

Experiments	Combination of succession measurements in the SPSA Algorithm		Controllable factors		Response variable	Number of iterations
	a_k	c_k	X_1^*	X_2^*	$(\widehat{C_{pk}})$	Position
1	1	1	375.230	44.139	20.984	500
2	2	1	375.230	44.138	20.984	500
3	3	1	375.096	45.019	21.054	500
4	4	1	382.716	46.016	21.238	185
5	5	1	375.165	44.561	21.021	500
6	1	2	375.214	44.142	20.984	500
7	2	2	375.215	44.141	20.984	500
8	3	2	375.077	45.045	21.055	500
9	4	2	382.759	46.025	21.239	183
10	5	2	375.149	44.568	21.021	500

Table 9 Results of the $\widehat{C_{pk}}$ evaluated by the SPSA Algorithm in the simulation process from 1 to 100 iterations in the DHA model reported by Gedi et al. [24]

Experiments	Combination of succession measurements in the SPSA Algorithm		Controllable factors		Response variable	Number of iterations
	a_k	c_k	X_1^*	X_2^*	$(\widehat{C_{pk}})$	Position
1	1	1	299.999	55.0053	4.655	49
2	2	1	299.999	55.0053	4.655	49
3	3	1	299.982	55.1135	4.680	54
4	4	1	299.972	55.1780	4.695	58
5	5	1	299.995	55.0298	4.661	50
6	1	2	299.999	55.0049	4.655	48
7	2	2	299.999	55.0049	4.655	48
8	3	2	299.984	55.1036	4.678	53
9	4	2	299.975	55.1637	4.692	56
10	5	2	299.995	55.0274	4.660	49

iterations possible. For example, experiment number 3, in which the combination of the succession measurements of succession $a_{k3} = \frac{a}{k}$ with $C_{k1} = \frac{c}{k^{\gamma}}$ is evaluated, the maximum value of the response variable $(\widehat{C_{pk}})$ that is obtained is of 5.69 units in 9 iterations, representing 0 (zero) ppm of defects in manufactured products.

The values obtained in each experiment in the iterative process from one to a hundred iterations of the simulation process of the model reported in Olguín [43], see page 103, model for industrial process 1 applying the (SPSA) Algorithm are shown graphically in Fig. 1.

Then, in Table 3, the results obtained for the iterative simulation process from 1 to 200 iterations of the model reported in Olguín [43], page 103 industrial process 1 are presented, as well as the maximum values found in the actual process capability index. $(\widehat{C_{pk}})$.

According to the results obtained in Table 3, experiments 3, 8, and 9 are maintained with acceptable values of the actual process capability index $\widehat{C_{pk}}$ as well as in the simulation process of 1–100 iterations. However, it is added experiments 5 and 10, which show acceptable values for the purpose of this research work, which are: obtain maximum values of $\widehat{C_{pk}} \geq 1.33$ in the least possible number of iterations, noting that experiments 3 and 9 still show the same maximum value of the response variable in the same number of iterations; Experiment number 8 increased the maximum value of the response variable from 5.86 units in 24 iterations to 23.62 units in 169 iterations. For experiments 5 and 10 acceptable values of 10.76 and 8.41 units were found in iterations 198 and 176 respectively.

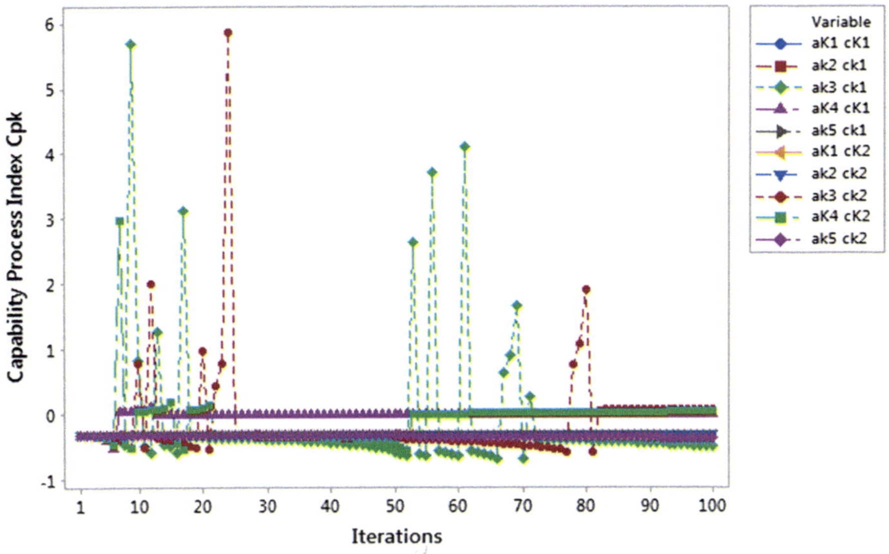

Fig. 1 Graphical behavior of ($\widehat{C_{pk}}$) in the simulation process from 1 to 100 iterations in model reported by (Olguín, et al. [43], page 103 industrial process 1)

The values obtained in each experiment in the iterative process of simulation of the model reported by Olguín [43], page 103 industrial process 1, from 1 to 200 iterations are shown graphically in Fig. 2.

Similarly, the results for the simulation process from 1 to 500 iterations and the maximum values of the actual process capability index $\widehat{C_{pk}}$ found are shown in Table 4.

The results for the iterative process of 1–500 iterations are shown in Table 4, where it can be observed that experiments 3, 5, 8, 9, and 10 are still the experiments in which the acceptable values of the actual Process Capability Index $\widehat{\boldsymbol{C}_{\boldsymbol{pk}}} \geq \mathbf{1.33}$ in the least number of iterations or possible experiments are found. On the other hand, in experiments 3 and 9 the value of the response variable is not increased, the maximum value is maintained (5.69 and 2.97 respectively) in the first iterations (9 and 7 respectively); in the rest of the experiments (5, 8, and 10) the maximum value of the response variable found increases (48.21, 78.36 and 41.28 units respectively) as well as the iteration number in which they are obtained (449, 259, and 250) iterations respectively).

The values obtained in each experiment in the iterative process of 1–500 iterations by means of the experimental simulation process of the model reported by Olguín [43], page 103 industrial process 1, are shown graphically in Fig. 3.

On the other hand, for the experimental simulation analysis applied to the regression models reported by Gedi et al. [24], the results for the variable YIELD are shown in Table 5, considering an iterative process of 1–100 iterations and the maximum values of the actual Process Capability Index $\widehat{C_{pk}}$ found.

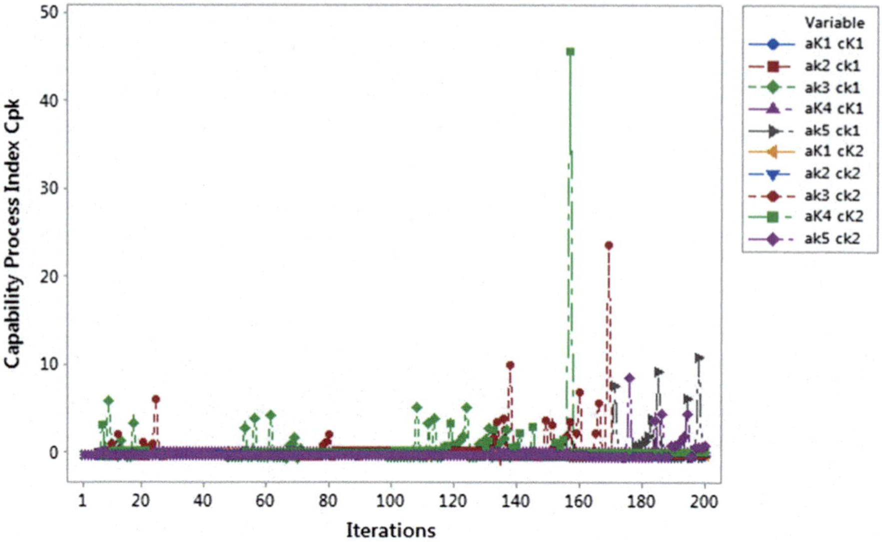

Fig. 2 Graphical behavior of $(\widehat{C_{pk}})$ in the simulation process from 1 to 200 iterations in model reported by (Olguín, et al. [43], page 103 industrial process 1)

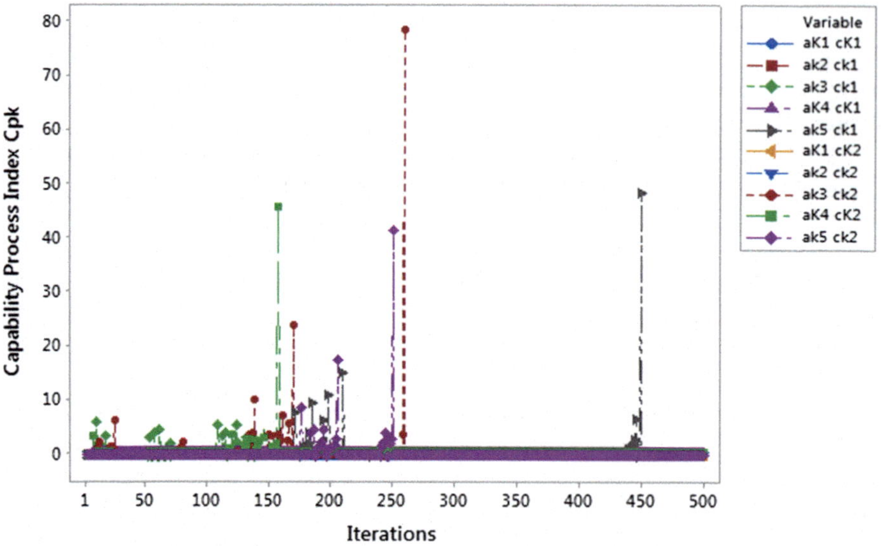

Fig. 3 Graphical behavior of $(\widehat{C_{pk}})$ in the simulation process from 1 to 500 iterations in model reported by (Olguín, et al. [43], page 103 industrial process 1)

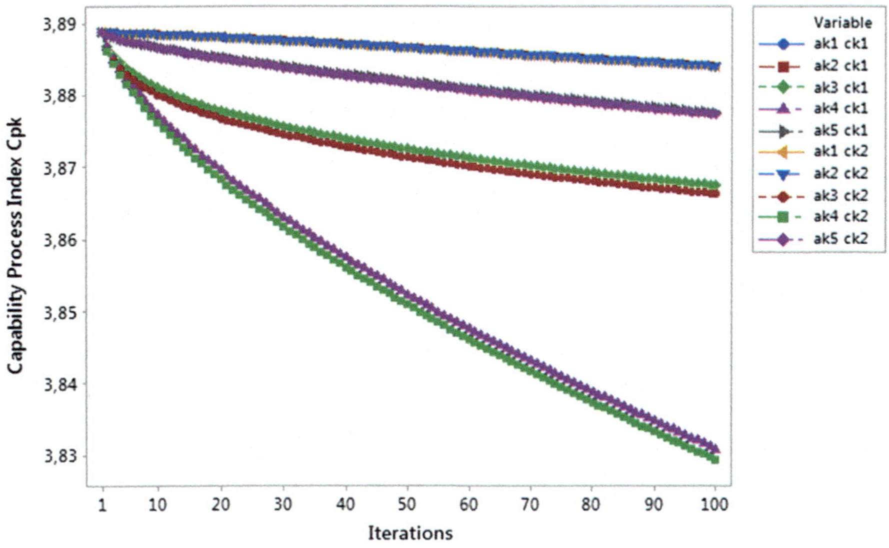

Fig. 4 Graphic behavior of $(\widehat{C_{pk}})$ in the simulation process from 1 to 100 iterations in the YIELD model reported by Gedi et al. [24]

According to the results shown in Table 5, it can be observed that the ten experiments yield the same maximum value of $\widehat{C_{pk}}$ of 3,889 units in the first iteration; taking into account that the acceptable values for the objective of this research work is to obtain values of de $\widehat{C_{pk}} \geq 1.33$ in the least number of iterations possible, in the ten experiments, the SPSA Algorithm manages to find this value in the first iteration.

It should be noted that the iterative processes of experimental simulation were performed for 200 and 500 iterations for the models reported by Gedi et al. [24], however, no significant improvements were obtained in the results obtained. The maximum values of $\widehat{C_{pk}}$ were always found in the first iteration in each of the experiments carried out.

The behavior of the values obtained for the simulation process of the YIELD model reported by Gedi et al. [24] in each of the ten simulation experiments performed can be seen in Fig. 4, which shows in the first 100 iterations the descending behavior of each of the experiments performed from the first iteration.

Next, Table 6 shows the maximum values of the actual process capability index $(\widehat{C_{pk}})$ found for the iterative process of simulation of the EPA model reported by Gedi et al. [24] from 1 to 100 iterations.

The results of Table 6 are the experiments performed on the EPA model reported by Gedi et al. [24] that show that the ten experiments performed find acceptable values $(\widehat{C_{pk}} \geq 1, 33)$, however, experiments 3, 4, 8, and 9, have the maximum values of $\widehat{C_{pk}}$ of 20.891, 21.169, and 20.897. 21.173 units respectively. These results were achieved in the 100th iteration of the simulation process.

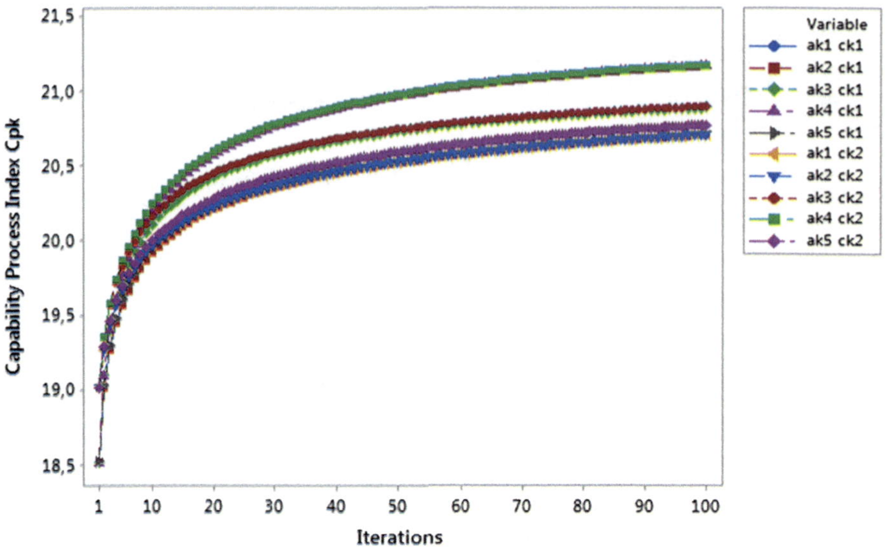

Fig. 5 Graphic behavior of $(\widehat{C_{pk}})$ in the simulation process from 1 to 100 iterations in the EPA model reported by Gedi et al. [24]

Figure 5 shows the behavior of the values obtained for the EPA simulation process in each of the ten experiments performed, in which, the ascending behavior presented by each of the experiments performed from the first iteration is shown in the first 100 iterations.

Next, Table 7 shows the results obtained for the iterative process of simulation of the EPA model from 1 to 200 iterations, as well as the maximum values found of the actual Process Capability Index $(\widehat{C_{pk}})$.

Table 7 shows that the results in each of the ten experiments performed, as well as in the simulation process of 1–100 iterations, the actual process capacity index $\widehat{C_{pk}}$ is maintained with acceptable values. Experiments 3, 4, 8 and 9 are still those that present the maximum values of $\widehat{C_{pk}}$ of 20.994, 21.238, 20.998, and 21.239 respectively. However, it is important to note that in experiments 4 and 9 are the maximum values of $\widehat{C_{pk}}$ in a smaller number of iterations

Figure 6 shows the values obtained in each experiment in the iterative process of simulation of the EPA model from one to two hundred iterations.

Next, Table 8 shows the maximum values obtained from the actual process capability index $(\widehat{C_{pk}})$ for the iterative process of simulation of the EPA model from 1 to 500 iterations.

Table 8 shows that the results in each of the ten experiments performed, as well as in the simulation process of 1–100 and of 1–200 iterations, the actual process capacity index $\widehat{C_{pk}}$ is maintained with acceptable values. Most of these maximum values occurred in the last iteration. It should be noted that experiments 3, 4, 8, and

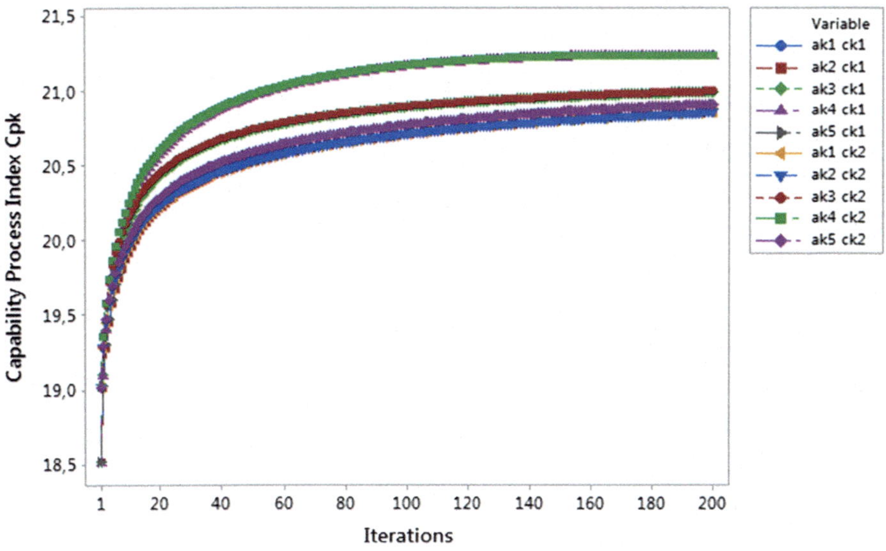

Fig. 6 Graphic behavior of $(\widehat{C_{pk}})$ in the simulation process from 1 to 200 iterations in the EPA model reported by Gedi et al. [24]

9 are again those with the maximum values of $\widehat{C_{pk}}$ of 21.054, 21.238, 21.055, and 21.239 respectively. However, it is important to note that in experiments 4 and 9 occurred the maximum values of $\widehat{C_{pk}}$ in a smaller number of iterations.

Figure 7 shows the behavior of the values obtained for the EPA simulation process in each of the ten simulation experiments performed with 1–500 iterations of the behavior presented by each of the experiments conducted.

Finally, the maximum values obtained from the actual process capability index $(\widehat{C_{pk}})$ for the iterative process of simulation of the DHA model from 1 to 100 iterations are shown in Table 9.

The results of Table 9 show that in each of the ten experiments carried out, the acceptable values of the actual process capacity index $\widehat{C_{pk}} \geq 1.33$ are maintained in the smallest possible number of iterations, this for the DHA model reported by Gedi et al. to the [24]. Experiments 3, 4, 8, and 9 have the highest values of $\widehat{C_{pk}}$ of 4,680, 4,695, 4,678, and 4,692 respectively. Therefore, it is important to note that in experiments 4 and 9 are the maximum values of $\widehat{C_{pk}}$ with 58 and 56 iterations respectively. However, there are values of $\widehat{C_{pk}}$ that have 48 iterations in experiments 6 and 7 and meet the acceptable value of $\widehat{C_{pk}} \geq 1.33$.

On the other hand, iterative simulation processes of 200 and 500 iterations were performed in this model. However, no significant improvements were obtained in the results attained. The maximum values for the response variable $\widehat{C_{pk}}$ were always found in the first iteration in each of the experiments carried out.

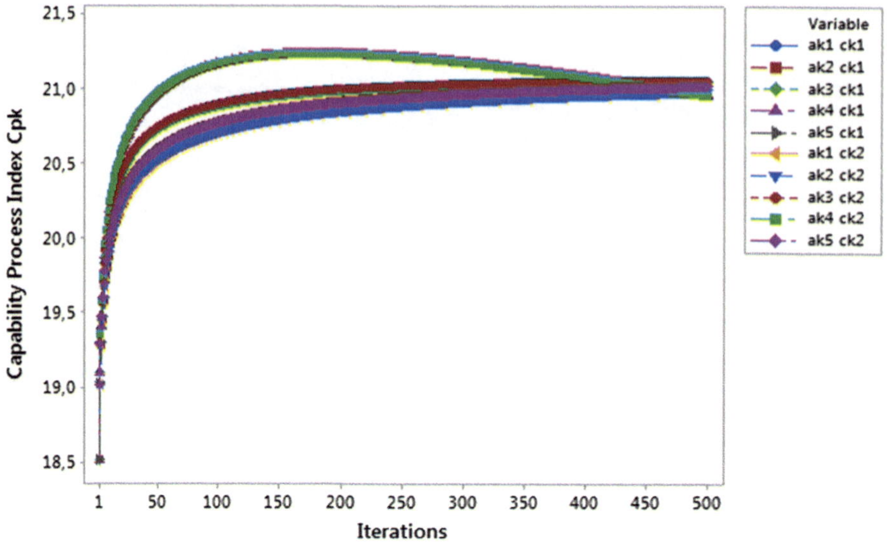

Fig. 7 Graphic behavior of ($\widehat{C_{pk}}$) in the simulation process from 1 to 500 iterations in the EPA model reported by Gedi et al. [24]

The behavior of the values obtained for the simulation process of the DHA process in each of the ten experiments carried out is shown in Fig. 8, which in the first 40 iterations maintains an ascending behavior, and as it progresses in the iterative process, the values of the $\widehat{C_{pk}}$ descend and manage to stabilize.

Taking into account that the main objective of this research is to find the best value for the response variable $\widehat{C_{pk}}$ in the least number of iterations possible, then it is assumed that the acceptable values of $\widehat{C_{pk}} \geq 1.33$ are considered as acceptable levels in standards of quality, but values of $\widehat{C_{pk}} \geq 2.00$ are considered as world class standards.

Of the four simulation models analyzed, the best results found in this chapter are listed below.

For the simulation model number 1, the following maximum values were obtained for the $\widehat{C_{pk}}$:

- In experiment 3, a value of $\widehat{C_{pk}} = 5.6935$ was obtained in iteration number 9.
- In experiment 8, a value of $\widehat{C_{pk}} = 5.8698$ was obtained in iteration number 24.
- In experiment 9, a value of $\widehat{C_{pk}} = 2.9784$ was obtained in iteration number 7.

For simulation model number 2, it was found that the ten experiments yielded their maximum value of $\widehat{C_{pk}}$ in the first iteration, this value being the same for each of the experiments:

- In experiments from 1 to 10, a value of $\widehat{C_{pk}} = 3.889$ was obtained in iteration number 1.

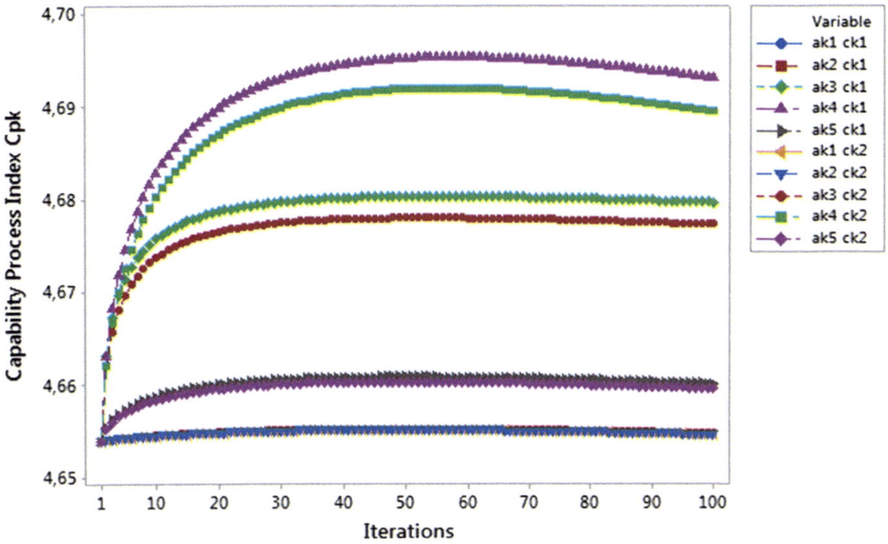

Fig. 8 Graphic behavior of $(\widehat{C_{pk}})$ in the simulation process from 1 to 100 iterations in the DHA model reported by Gedi et al. [24]

For simulation model number 3, the following maximum values were found for the $\widehat{C_{pk}}$:

- In experiment 3, a value of $\widehat{C_{pk}} = 20.891$ is obtained in iteration number 100.
- In experiment 4, a value of $\widehat{C_{pk}} = 21.169$ is obtained in iteration number 100.
- In experiment 8, a value of $\widehat{C_{pk}} = 20.897$ is obtained in iteration number 100.
- In experiment 9, a value of $\widehat{C_{pk}} = 21.173$ is obtained in iteration number 100.

And finally, for simulation model number 4, the following maximum values were found for the response variable:

- In experiment 3, a value of $\widehat{C_{pk}} = 4.680$ is obtained in iteration number 54.
- In experiment 4, a value of $\widehat{C_{pk}} = 4.695$ is obtained in iteration number 58.
- In experiment 8, a value of $\widehat{C_{pk}} = 4.678$ is obtained in iteration number 53.
- In experiment 9, a value of $\widehat{C_{pk}} = 4.692$ is obtained in iteration number 56.

4 Conclusions

In this experimental simulation work, 10 combinations of the a_k and c_k succession measurements found in the literature were evaluated as part of the efficiency of the SPSA Algorithm, so they were used in four case studies, in which the objective is to maximize the actual process capability index $\widehat{C_{pk}}$ in the smallest number of iterations.

Each of the successions evaluated by case study was simulated to 100, 200 and 500 iterations to determine the optimal solutions in each of the study processes, this, in order to identify if any of these experiments offered better results than others, meaning better results those equal to 1.33. The results showed in the four study cases analyzed by means of experimental simulation that at least in the four second order regression models, the SPSA Algorithm is able to optimize response variables such as the actual process capability index $\widehat{C_{pk}}$, not only for analytically modeled systems, but also for systems in which only simulated values of the objective function are available as reported by Miranda and Del Castillo [41].

So it can be concluded that there is a difference in the solution that the SPSA Algorithm can achieve depending on the best combination in the succession measurements a_k and c_k that are used to perform the analysis. For the four case studies, the succession measurements that meet the two main objectives of this simulation work is the experiment number 3 ($a_{k3} = a/k$ and $c_{k1} = c/k^{\gamma}$) which achieves acceptable values for the variable of response in the least number of iterations in the 4 models analyzed.

In addition to experiment number 3, the experiments 8 ($a_{k3} = a/k$ y $c_{k1} = c/(k+1)^{\gamma}$) and 9 ($a_{k3} = a/k^{2/3}$ y $c_{k1} = c/(k+1)^{\gamma}$) appear in the maximum values for the response variable $\widehat{C_{pk}}$, although with a greater number of iterations in the 4 simulation models evaluated which permit us to conclude that these experiments are also useful to maximize the actual process capability index $\widehat{C_{pk}}$.

Future Works

Future research work will evaluate other Process Capability Indexes such as the Taguchi index (Cpm), the Dual Response Index (DR), and the Signal-to-Noise-Ratio for process capability (SNR).

Acknowledgements This work was supported by the research project approved at the 18th Internal Call 580 for Research Projects by UABC, with number 485. The researcher J. C. C. G. was supported for his postgraduate studies at Ph.D. level by CONACyT. Thanks to PRODEP (Professional Development Program for Professors) for supporting the new generations and for innovating the application of knowledge.

Compliance with Ethical Standards

Conflict of Interest The authors declare that they have no conflict of interest.

References

1. Abdolshah, M., Ismail, Y.B., Yusuff, R.M., Hong, T.S.: Process capability analysis using Monte Carlo simulation. In: International Conference on Information Management and Engineering, 2009. ICIME'09, pp. 335–339. IEEE (2009)
2. Abdulsadda, A.T., Iqbal, K.: An improved SPSA algorithm for system identification using fuzzy rules for training neural networks. Int. J. Autom. Comput. **8**(3), 333–339 (2011). https://doi.org/10.1007/s11633-011-0589-x

3. Ahmad, S., Abdollahian, M., Zeephongsekul, P.: Process capability for a non-normal quality characteristics data. In: Fourth International Conference on Information Technology, 2007. ITNG'07, pp. 420–424. IEEE (2007)
4. Altaf, M.U., Heemink, A.W., Verlaan, M., Hoteit, I.: Simultaneous perturbation stochastic approximation for tidal models. Ocean Dyn. **61**(8), 1093–1105 (2011). https://doi.org/10.1007/s10236-011-0387-6
5. Andradóttir, S.: A stochastic approximation algorithm with varying bounds. Oper. Res. **43**(6), 1037–1048 (1995)
6. Aslam, M., Wu, C.W., Azam, M., Jun, C.H.: Variable sampling inspection for resubmitted lots based on process capability index Cpk for normally distributed items. Appl. Math. Model. **37**(3), 667–675 (2013)
7. Bai, B., Ju, F., Zhang, J.: Study on process capability index improvement model. In: 2011 2nd International Conference on Artificial Intelligence, Management Science and Electronic Commerce (AIMSEC), pp. 1543–1545. IEEE (2011)
8. Baltcheva, I., Cristea, S., Vázquez-Abad, F.J., De Prada, C.: Simultaneous perturbation stochastic approximation for real-time optimization of model predictive control. In: Proceedings of the First Industrial Simulation Conference (ISC 2003), Valencia, Spain, June 2003, pp. 533–537 (global optimization for model predictive control) (2003)
9. Bangerth, W., Klie, H., Matossian, V., Parashar, M., Wheeler, M.F.: An autonomic reservoir framework for the stochastic optimization of well placement. Cluster Comput. **8**(4), 255–269 (2005)
10. Bartkute, V., Sakalauskas, L.: Simultaneous perturbation stochastic approximation of nonsmooth functions. Eur. J. Oper. Res. **181**, 1174–1188 (2005)
11. Brittanti, S., Lovera, M., Moiraghi, L.: Application of non-normal process capability indices to semiconductor quality control. IEEE Trans. Semicond. Manuf. **11**(2), 296–303 (1998)
12. Broadie, M., Cicek, D., Zeevi, A.: General bounds and finite-time improvement for the Kiefer-Wolfowitz stochastic approximation algorithm. Oper. Res. **59**(5), 1211–1224 (2011)
13. Brooks, O.: Solving discrete resource allocation problems using the Simultaneous Perturbation Stochastic Approximation (SPSA) algorithm. In: SpringSim, no 3, pp. 55–62 (2007)
14. Cao, X.: Preliminary results on non-bernoulli distribution of perturbations for simultaneous perturbation stochastic approximation. In: Proceedings of the 2011 American Control Conference, pp. 2669–2670 (2011). https://doi.org/10.1109/acc.2011.5991550
15. Chin, D.C.: Comparative study of stochastic algorithms for system optimization based on gradient approximations. IEEE Trans. Syst. Man Cybern. Part B: Cybern. **27**(2), 244–249 (1997)
16. Cipriani, E., Florian, M., Mahut, M., Nigro, M.: A gradient approximation approach for adjusting temporal origin–destination matrices. Transp. Res. Part C: Emerg. Technol. **19**(2), 270–282 (2011). https://doi.org/10.1016/j.trc.2010.05.013
17. de Felipe, D., Benedito, E.: A review of univariate and multivariate process capability indices. Int. J. Adv. Manuf. Technol. **92**(5–8), 1687–1705 (2017)
18. Dellino, G., Kleijnen, J.P., Meloni, C.: Robust optimization in simulation: Taguchi and Krige combined. INFORMS J. Comput. **24**(3), 471–484 (2012)
19. Ding, S., Xia, N., Wang, P., Li, S., Ou, Y.: Optimization algorithm based on SPSA in multi-channel multi-radio wireless monitoring network. In: 2015 International Conference on Cyber-Enabled Distributed Computing and Knowledge Discovery, pp. 517–524 (2015). https://doi.org/10.1109/cyberc.2015.84
20. Ebadi, M., Shahriari, H.: A process capability index for simple linear profile. Int. J. Adv. Manuf. Technol. **64**(5–8), 857–865 (2013)
21. Finck, S., Beyer, H.G.: Performance analysis of the simultaneous perturbation stochastic approximation algorithm on the noisy sphere model. Theoret. Comput. Sci. **419**, 50–72 (2012)
22. Fu, M.C., Ho, Y.C.: Using perturbation analysis for gradient estimation, averaging and updating in a stochastic approximation algorithm. In: Proceedings of the 20th Conference on Winter Simulation, pp. 509–517 (1988)

23. García, M.R.: Algoritmo de Aproximaciones Estocásticas para el Mejoramiento del Índice de Capacidad Real de Procesos CPK. Tesis de Doctorado en Ingeniería Industrial, Instituto Tecnológico de Cd. Juárez (2000)
24. Gedi, M.A., Bakar, J., Mariod, A.A.: Optimization of supercritical carbon dioxide (CO_2) extraction of sardine (Sardinella lemuru Bleeker) oil using response surface methodology (RSM). Grasas Aceites **66**(2), 074 (2015)
25. Haghighat Sefat, M., Muradov, K.M., Elsheikh, A.H., Davies, D.R.: Proactive optimization of intelligent-well production using stochastic gradient-based algorithms. SPE Reserv. Eval. Eng. **19**(02), 239–252 (2016)
26. Hosseinifard, S.Z., Abbasi, B., Ahmad, S., Abdollahian, M.: A transformation technique to estimate the process capability index for non-normal processes. Int. J. Adv. Manuf. Technol. **40**(5–6), 512 (2009)
27. Ito, K., Dhaene, T.: Adaptive initial step size selection for Simultaneous Perturbation Stochastic Approximation. SpringerPlus **5**(1), 200 (2016). https://doi.org/10.1186/s40064-016-1823-3
28. Jalali, H., Nieuwenhuyse, I.V.: Simulation optimization in inventory replenishment: a classification. IIE Trans. **47**(11), 1217–1235 (2015)
29. Jianfeng, Y.: Process capability evaluation in skew distributed processes based on SWV method. In: International Conference on Wireless Communications, Networking and Mobile Computing, 2007. WiCom 2007, pp. 5919–5922. IEEE (2007)
30. Kane, V.E.: Process capability indices. J. Qual. technol. **18**(1), 41–52 (1986)
31. Kashif, M., Aslam, M., Jun, C.H., Al-Marshadi, A.H., Rao, G.S.: The efficacy of process capability indices using median absolute deviation and their bootstrap confidence intervals. Arab. J. Sci. Eng. **42**(11), 4941–4955 (2017)
32. Khong, S.Z., Tan, Y., Manzie, C., Nešić, D.: Extremum seeking of dynamical systems via gradient descent and stochastic approximation methods. Automatica **56**, 44–52 (2015)
33. Kotz, S., Johnson, N.L. Process capability indices—a review, 1992–2000. J. Qual. Technol. **34**(1), 2–19 (2002)
34. L'Ecuyer, P., Giroux, N., Glynn, P.W.: Stochastic optimization by simulation: numerical experiments with the M/M/1 queue in steady-state. Manage. Sci. **40**(10), 1245–1261 (1994)
35. Lee, K.K., Ro, Y.C., Han, S.H.: Tolerance optimization of a lower arm by using genetic algorithm and process capability index. Int. J. Precis. Eng. Manuf. **15**(6), 1001–1007 (2014)
36. Lin, C.J., Pearn, W.L.: Process selection for higher production yield based on capability index Spk. Qual. Reliab. Eng. Int. **26**(3), 247–258 (2010)
37. Lofthouse, T.: The Taguchi loss function. Work Study **48**(6), 218–223 (1999)
38. Maryak, J.L., Chin, D.C.: Global random optimization by simultaneous perturbation stochastic approximation. IEEE Trans. Autom. Control **53**, 780–783 (2008)
39. McClary, D.W., Syrotiuk, V.R., Kulahci, M.: Steepest-ascent constrained simultaneous perturbation for multiobjective optimization. ACM Trans. Model. Comput. Simul. **21**(1), 1–22 (2010). https://doi.org/10.1145/1870085.1870087
40. Méndez-González, L.C., Rodríguez-Picón, L.A., Quezada-Carreón, A.E., Romero-López, R., Garcia, V.: Process capability index for AC transformer under electrical harmonics. Electr. Eng. **100**(2), 347–353 (2018)
41. Miranda, A.K., Del Castillo, E.: Robust parameter design optimization of simulation experiments using stochastic perturbation methods. J. Oper. Res. Soc. **62**(1), 198–205 (2011)
42. Montgomery, D.C.: Design and Analysis of Experiments. Wiley (2008)
43. Olguín, J.: Algoritmo de aproximaciones estocásticas para la optimización de procesos industriales. Ingeniería e investigación **31**(3), 100–111 (2011)
44. Parnianifard, A., Azfanizam, A.S., Ariffin, M.K.A., Ismail, M.I.S.: An overview on robust design hybrid metamodeling: advanced methodology in process optimization under uncertainty. Int. J. Ind. Eng. Comput. 1–32 (2017)
45. Pearn, W.L., Shiau, J.J., Tai, Y.T., Li, M.Y.: Capability assessment for processes with multiple characteristics: a generalization of the popular index Cpk. Qual. Reliab. Eng. Int. **27**(8), 1119–1129 (2011)

46. Prashanth, L.A., Bhatnagar, S., Fu, M., Marcus, S.: Adaptive system optimization using random directions stochastic approximation. IEEE Trans. Autom. Control **62**(5), 2223–2238 (2017)
47. Radenković, M.S., Stanković, M.S., Stanković, S.S.: On stochastic extremum seeking via adaptive perturbation-demodulation loop. J. Optim. Theory Appl. **179**(3), 1008–1024 (2018)
48. Ramakrishnan, B., Sandborn, P., Pecht, M.: Process capability indices and product reliability. Microelectron. Reliab. **41**(12), 2067–2070 (2001)
49. Reardon, B.E., Lloyd, J.M., Perel, R.Y.: Tuning missile guidance and control algorithms using simultaneous perturbation stochastic approximation. Johns Hopkins APL Tech. Digest **29**(1), 85–100 (2010)
50. Rodriguez, R.N.: Discussion-process capability indices—a review, 1992–2000. J. Qual. Technol. **34**(1), 28–31 (2002)
51. Ros-Roca, X., Montero, L., Schneck, A., Barceló, J.: Investigating the performance of SPSA in simulation-optimization approaches to transportation problems. Transp. Res. Procedia **34**, 83–90 (2018)
52. Sadegh, P., & Spall, J.C.: Optimal random perturbations for stochastic approximation using a simultaneous perturbation gradient approximation. In: Proceedings of the 1997 American Control Conference (Cat. No. 97CH36041), vol. 6, pp. 3582–3586 (1997). https://doi.org/10.1109/acc.1997.609490
53. Seyedpoor, S.M., Salajegheh, J., Salajegheh, E., Gholizadeh, S.: Optimal design of arch dams subjected to earthquake loading by a combination of simultaneous perturbation stochastic approximation and particle swarm algorithms. Appl. Soft Comput. **11**(1), 39–48 (2011). https://doi.org/10.1016/j.asoc.2009.10.014
54. Shiau, J.J.H., Hung, H.N., Chiang, C.T.: A note on Bayesian estimation of process capability indices. Stat. probab. lett. **45**(3), 215-224 (1999)
55. Sidorov, K.A., Richmond, S., Marshall, D.: An efficient stochastic approach to groupwise non-rigid image registration. In: 2009 IEEE Conference on Computer Vision and Pattern Recognition, pp. 2208–2213 (2009). https://doi.org/10.1109/cvpr.2009.5206516
56. Śliwiński, P., Wachel, P.: A simple model for on-sensor phase-detection autofocusing algorithm. J. Comput. Commun. **1**(06), 11 (2013)
57. Spall, J.C.: Multivariate stochastic approximation using a simultaneous perturbation gradient approximation. IEEE Trans. Autom. Control **37**(3), 332–341 (1992)
58. Spall, J.C.: Implementation of the simultaneous perturbation algorithm for stochastic optimization. IEEE Trans. Aerosp. Electron. Syst. **34**(3), 817–823 (1998)
59. Spall, J.C.: An overview of the simultaneous perturbation method for efficient optimization. Johns Hopkins APL Tech. Digest **19**(4), 482–492 (1998)
60. Spall, J.C.: A stochastic approximation algorithm for large-dimensional systems in the Kiefer-Wolfowitz setting. In: Proceedings of the 27th IEEE Conference on Decision and Control, 1988, pp. 1544–1548. IEEE (1988)
61. Spall, J.C.: Stochastic optimization. In: Handbook of Computational Statistics, pp. 173–201. Springer, Berlin, Heidelberg (2012)
62. Stoumbos, Z.G.: Process capability indices: overview and extensions. Nonlinear Anal.: Real World Appl. **3**(2), 191–210 (2002)
63. Tang, S., Wang, F.Y.: A PCI-based evaluation method for level of services for traffic operational systems. IEEE Trans. Intell. Transp. Syst. **7**(4), 494–499 (2006)
64. Venkatesh, Y.V., Kassim, A.A., Zonoobi, D.: Medical Image Reconstruction from Sparse Samples Using Simultaneous Perturbation Stochastic Optimization, pp. 3369–3372. National University of Singapore, Singapore (2010)
65. Wang, L.-F., Shi, L.-Y.: Simulation optimization: a review on theory and applications. Acta Automatica Sinica **39**(11), 1957–1968 (2013)
66. Wang, Q.: Stochastic approximation for discrete optimization of noisy loss measurements. In: 2011 45th Annual Conference on Information Sciences and Systems, pp. 1–4 (2011)
67. Wu, C.W., Pearn, W.L., Kotz, S.: An overview of theory and practice on process capability indices for quality assurance. Int. J. Prod. Econ. **117**(2), 338–359 (2009)

68. Yum, B.J., Kim, K.W.: A bibliography of the literature on process capability indices: 2000–2009. Qual. Reliab. Eng. Int. **27**(3), 251–268 (2011)

An Optimization Vehicle Routing Problem Approached by Bio-inspired Algorithms—A Real Case Study

Eliandis Matos, Fernando Gaxiola, Luis Carlos González-Gurrola and Alain Manzo-Martinez

Abstract In this study, an optimization with genetic algorithm and multi-objective genetic algorithm for vehicle transportation route is performed. The objective function used in the genetic algorithm is calculated using an equation with the sum of the values of the total distance and the number of active vehicles. For the multi-objective genetic algorithm the values are considered as separated objectives. The proposed approach is applied to the case of transportation routes of workers for the industrial plant "JABIL Circuit" in Chihuahua City (Mexico) for evaluating its efficiency. The results presented in this paper are state-of-the-art, since in comparison with the actual route planning, the number of vehicles in the original fleet is considerably reduced, and the total distance traveled by all routes also is minimized; also the quality of service for the workers is maintained (transportation time).

1 Introduction

In this study, the optimization of a transportation problem for workers of the industry utilizing the genetic algorithms (GA) and multi-objective genetic algorithm (MOGA) are proposed. The vehicle route problem (VRP) is a complex problem in terms of computation, besides the solution using traditional methods present a high cost and this can be not the optimal solution [1, 2].

The genetic algorithm is a method used for optimization and search, therefore this method is a great candidate for the problem of this work; besides, in the literature, this algorithm obtained optimal solutions and in acceptable time for a diversity of problems. In this work, the genetic algorithm is employed to look for the optimization routes and the minimum vehicles to utilize for the routes [3].

The main contribution of this work are the development and comparison of the results between the genetic algorithm and the multi-objective genetic algorithm. Currently, in the area of vehicle routing problems (VRP) exists many research projects [4, 5, 6]. For this work, the more significant projects in literature are the following:

E. Matos · F. Gaxiola (✉) · L. C. González-Gurrola · A. Manzo-Martinez
Autonomous University of Chihuahua, Chihuahua, Mexico
e-mail: lgaxiola@uach.mx

M. A. Sanchez et al. (eds.), *Applied Decision-Making*, Studies in Systems, Decision and Control 209, https://doi.org/10.1007/978-3-030-17985-4_2

Baker and Ayechew [7], explain that the best results for VRP has been obtained utilizing the Tabú search or the Simulating Annealing; the use of genetic algorithm presented good results but not better than the tabú search, therefore, the hybridization of genetic algorithm (GA) with others heuristic methods with the aim to increase the quality of the results is applied, like the hybrid of GA with neighborhood search method.

Zuñiga and González [8], developed an Open Vehicle Routing Problem Capacitated with Distance constraint (OVRPCD) for personal transporting with balance; the Mixed Integer Lineal Programming (MILP) or the Greedy Randomized Adaptive Search Procedure (GRASP) is the method applied in the work, and the results presented are good solutions in almost the cases in a reduced execution time.

The use of genetic algorithm (GA) is widely applied for optimization problems, either working with basic GA or multiobjective GA; besides, the results with this methods are optimal and in many cases better than others methods [9].

Saravanan et al. [10], perform an optimization of the operations for a superficial rectified using a multiobjective genetic algorithm. The research present an optimization procedure for the conditions of trituration, wheel velocity, velocity of the work piece, profundity of the bandage and lead of bandage utilizing a model of function multiobjective with a weighted approach for the superficial rectified. The results obtained with GA are compared and better than the ones for quadratic programming method (QA).

The next section presents a background about optimization of vehicle routing problems in literature, and basic concepts of vehicle routing problems and bio-inspired algorithms. Section 3 explains the proposed method and the problem description. Section 4 presents the simulation results for the proposed method. Finally, in Sect. 5 some conclusions are presented.

2 Background

In this section we introduce basic concepts that would help contextualize this study.

2.1 Vehicle Routing Problem

The Vehicle Routing Problem (VRP) is very well-known within the field of Operations Research (OR). It is used to model the management of a fleet of vehicles that need to deliver goods to a set of customers. Generally speaking, it is represented as a bi-objective optimization problem, where its main objectives consider the total sum of distances of all vehicles and the number of vehicles that are to be used. Given the possible arrangements of routes based on the number customers to visit finding the optimal solution rapidly becomes impractical for cases motivated in the real-world [11, 12, 13]. Theorically speaking, the VRP is a NP-complete combinatorial problem, meaning that no algorithm is known to optimally solve this problem in polynomial time [14, 15, 16].

2.1.1 Vehicle Routing Problem with the Objective of Route Balance

The vehicle routing problem with Route Balance (VRPRB) is a variant of the classical VRP, in which, besides to minimize the total distance of all vehicles a balance in the cargo (load/freight) is sought [17, 18]. Adding the cargo balance as an objective allows to introduce certain grade of justice to the optimal solution without disregarding the economic aspect of the problem [19].

2.2 Bio-Inspired Algorithms

The bio-inspired algorithms are heuristics that emulate the strategic actions of the nature, this approach is conceived in base that many biologic process can be formulated like processes of restricted optimization. These algorithms utilizes many random decisions that allow to classify them as a special class of Randomized Algorithms (RA). The application of a bio-inspired algorithm to solve a problem consist in choosing an adequate representation of the problem, aformulate a fitness function to evaluate the quality of the solutions, and define the operators to use to produce a new set of solutions [20].

2.2.1 Genetic Algorithm

The Genetic Algorithm (GA) is a method of optimization and search inspired by an abstraction of the Darwin evolutionary theory, based in the principles of genetic and natural selection. The GA allows that many individuals of a population evolve based on specific selection rules, to an state of maximization or minimization of the objective function. The final result consists in the survival of the fittest individuals which represent the best found solution for the problem in hand.

The Genetic Algorithm can be organized in several procedures: representation, evaluation of the objective function (FO) or fitness calculation, initialization, selection, recombination (recombination and mutation) and termination [3]. Some important concepts of GA are explained next.

- ***Parents selection***: This process is used to identify individuals to reproduce (recombinate or crossover) and generate a new individual for the next generation. Exists many techniques for selection, however, usually, a process of natural selection is simulated, in which the stronger individuals (best objective function) are utilized; this manner increases the probability of descendants (offspring) to have good characteristics.

- ***Recombination or crossover***: The operators of crossover is an important part of the process of the GA since these operator simulate the reproduction of two individuals and allow the descendants to inherit characteristics of the progenitors.
- ***Mutation***: The mutation is the process of creating a new individual by making changes of some characteristics of a solution (chromosome). This procedure have as its main objective to increase the probability of search for a global optimum y not allow stagnation in a local optimal. Both, crossover and mutation operators are applied using a predefined probability.

2.2.2 Multi-objective Genetic Algorithm (MO-GA)

For Real-World applications, the optimization problems involve two or more objectives. The objectives to optimize can be in conflict and this make it specially hard to evaluate both via an aggregate function [21, 22, 23]. Also, for these cases, an extreme solution cannot provide an optimal solution for both objectives at the same time, i.e. the optimal solution obtained for one objective could not be optimal for other objective. Therefore, the Multi-objective Genetic Algorithm aims to find a set of individuals that represent good solutions of all the objectives and presents compensations between the different objectives. Basically, a multi-objective optimization problem work in the search space to find optimal solutions for two or more objectives [24].

2.3 Overview of Related Works

In Vaira [3] the genetic operators applied the approach of destruction and reconstruction used in the Large Neighborhood Search algorithm (LNS), in which the randomized insertion heuristic process used is replaced with a reconstruction method. New operators of crossover based in the search of common parts in the principal solutions for the generation of descendant is applied; this new approach produce solutions in lower time and with the same or better results than another Genetic Algorithms. Besides, the proposed operators are not designed for only one kind of specific problem and can be applied to distinct problems. The characteristics of the operators in the prior study are considered a very attractive solution for a starting point or comparison for the work in this paper.

Lozano et al. [19], performed a comparison of the results obtained with a genetic algorithm using different objective functions for the vehicle routing problem with routing balanced (VRPRB). In the experiments, the 95% of the executions with the different objective functions (FO) are obtained results diverse without matter significantly the operators of mutation or crossover.

3 Proposed Method and Problem Description

The objective of this work is the design of genetic algorithms for the integrated problem of planning and balancing of transporting routes of workers for the industrial plant named “JABIL Circuit”.

The transportation service is provided to 643 workers in a total of 363 “demand points” distributed through the city of Chihuahua, Mexico. The plant currently employs 27 vehicles (buses) to fulfill the current demand. All vehicles have a homogenous capacity of 40 workers.

For this study, the design of the structure of the chromosome includes a set of vehicles that in turn have incorporated the routes and the order through which they must pas. Furthermore, some restrictions are established, like that the capacity of each vehicle in the route must not be surpassed, given that at each point there is no demand above their capacity and they will only go through a single point once.

The database consists of 364 points for routes including the location of the factory. The representation of the chromosome is shown in Fig. 1.

Each chromosome is integrated for 34 blocks of 364 genes adding up in total of 12,376 genes. The firsts genes of each block of the chromosome represent the activation of each vehicle (1 = active, 0 = inactive); the remaining 363 genes represent the points (bus stops) where the bus must pass. For example, the routes of the chromosome in Fig. 1 is interpreted as follow:

- Vehicle 1: (Active) Factory—Point 123—Point 25—Point 3—Point 320—etc.
- Vehicle 2: (Inactive) Factory—Point 310—Point 40—Point 57—Point 21—etc.
- Vehicle 3: (Active) Factory—Point 241—Point 2—Point 10—Point 17—etc.
- Vehicle 4: (Active) Factory—Point 128—Point 5—Point 305—Point 19—etc.

In the genetic algorithm and multi-objective genetic algorithms the same chromosome is utilized.

The objective function of the genetic algorithm integrates in a mathematical equation (Eq. 1) the amount of active vehicles and the total distance of all routes (traveled routes).

$$\mathrm{FO} = \left(\sum_{\mathrm{i=1}}^{\mathrm{v}} a(V_i) * 10^R\right) + \left(\sum_{\mathrm{i=1}}^{\mathrm{v}} a(V_i) * dist(V_i)\right) \tag{1}$$

where:

FO is the objective function.
v is the quantity of vehicles.

Fig. 1 Design of the chromosome

V_i is the vehicle–in the position i.
R is the amount of digits in the total distance.
$a()$ is the function used to determine the activation of the vehicle.
$dist()$ is the function used to calculate the distance traveled for a vehicle.

The evaluation process of each individual starts from a list of visited points that is initialize empty, then, the chromosome fitness is obtained by integrating the values of the number of vehicles to utilize in the routes, the collect total capacity in all points and the total distance traveled obtained for all vehicles in the routes.

The process of evaluation is explained as follow:

1. The chromosome is separated in 34 blocks, each block represents one vehicle with the set of demand points that it will visit (route).
2. For each block, the first gene is analyzed to identify if the vehicle is active or inactive (1 or 0); in case of being active, the variable of the number of vehicles is incremented.
3. If the vehicle is active, the next genes represent the points to visit, each point is visited as long as the point that is being selected has not yet been visited and the its demand plus the accumulated capacity of the vehicle at that moment is less or equal than total vehicle capacity. Each point visited is added to the list of visited points and the distance between the prior point and the current point is calculated (in case of being the first point, the distance is calculated with respect to the plant.
4. When all vehicles are analyzed the total capacity of the route is verified; if this restriction is achieved, the evaluation of the fitness is calculated with the Eq. 1, where the first part of the Eq. 1 is the product of the multiplication of the number of accumulated vehicles by 10^R, and the second part is the accumulated distance of the active vehicles for the selected points to visit.

In Multi-objective GA, the variation in the evaluation of the fitness is the use of two objectives, one of them is the number of vehicles (Eq. 2) and the second is the total distance (Eq. 3). In case of the total capacity is not achieved (no feasible), the values assigned are infinite in order to discard these individuals in the process of selection from parents and reinsertion for the next generation.

$$\mathrm{FO}_1 = \left(\sum_{\mathrm{i}=1}^{\mathrm{v}} a(V_i) * 10^R\right) \tag{2}$$

$$\mathrm{FO}_2 = \left(\sum_{\mathrm{i}=1}^{\mathrm{v}} a(V_i) * dist(V_i)\right) \tag{3}$$

The pseudo-code of the fitness evaluation of the genetic algorithm and multi-objective GA is presented in Appendix 1.

The parameters for the genetic algorithm are shown in Table 1.

The parameters for the multi-objective genetic algorithm are shown in Table 2.

Table 1 Parameters for the genetic algorithm (GA)

Parameters	GA
Individuals	200
Genes	12,376 (34 * 364)
Generations	2000
Selection	Stochastic uniform
Crossover	Two points (0.8)
Mutation	Uniform (0.05)

Table 2 Parameters for the multi-objective genetic algorithm (MOGA)

Parameters	MOGA
Individuals	200
Genes	12,376 (34 * 364)
Generations	2000
Selection	Tournament
Crossover	Two points (0.8)
Mutation	Adaptive feasible

4 Simulation Results

We performed experiments for the optimization of vehicle routing problem, specifically for a real database of points obtained of the plant JABIL Circuit. The number of vehicles currently in use is of 27 and their total traveled distance is 339,223.91 (m).

We present the obtained results of the experiments performed with genetic algorithm (GA) and multi-objective genetic algorithm (MOGA), these results are

Table 3 Results of the number of vehicles, total distance and execution time for the genetic algorithm

Test	Vehicles	Total distance	Execution (h)
1	17	121180.00	0.021145
2	17	123860.00	0.019545
3	17	135520.00	0.01921833
4	17	123940.00	0.019535
5	17	125130.00	0.01830833
6	17	121510.00	0.01933167
7	17	135950.00	0.01789
8	17	120960.00	0.023035

(continued)

Table 3 (continued)

Test	Vehicles	Total distance	Execution (h)
9	17	113030.00	0.02392167
10	17	132380.00	0.01894
11	17	127830.00	0.02220833
12	17	124240.00	0.02128333
13	17	127760.00	0.02204667
14	17	124980.00	0.01936833
15	17	141270.00	0.02140833
16	17	129430.00	0.026445
17	17	123960.00	0.02021833
18	17	126080.00	0.02403
19	17	119340.00	0.02091667
20	17	136560.00	0.021595
21	18	116320.00	0.021745
22	17	138750.00	0.02301833
23	17	123600.00	0.02256333
24	17	128690.00	0.01915667
25	17	136480.00	0.02122
26	17	125560.00	0.01942167
27	17	122730.00	0.021175
28	17	123310.00	0.02120167
29	17	126810.00	0.02102667
30	17	120800.00	0.02112
Min	17	113030.00	0.02392167
Max	18	116320.00	0.021745
Ave	17.0333	126598.6667	0.021

achieved with the parameters presented in Tables 1 and 2. We performed 30 runs for the algorithms.

In Table 3, we present the results obtained by the GA, we report performance of the execution time, quantity of vehicles and total distance. The minimum quantity of vehicles is of 17, total distance is 113,030 m and the average is of 17.03 and 126,598.67 m, respectively.

In Figs. 2 and 3, we show the behavior of the objective function (quantity of vehicles and total distance) in the executions of 30 runs for the GA.

In Table 4, we present the results obtained by MOGA when measuring the execution time, number of vehicles and total distance. The minimum quantity of vehicles is of 17 in a total distance of 104,930 m, the average is of 18.27 and 110,677.8 m, respectively.

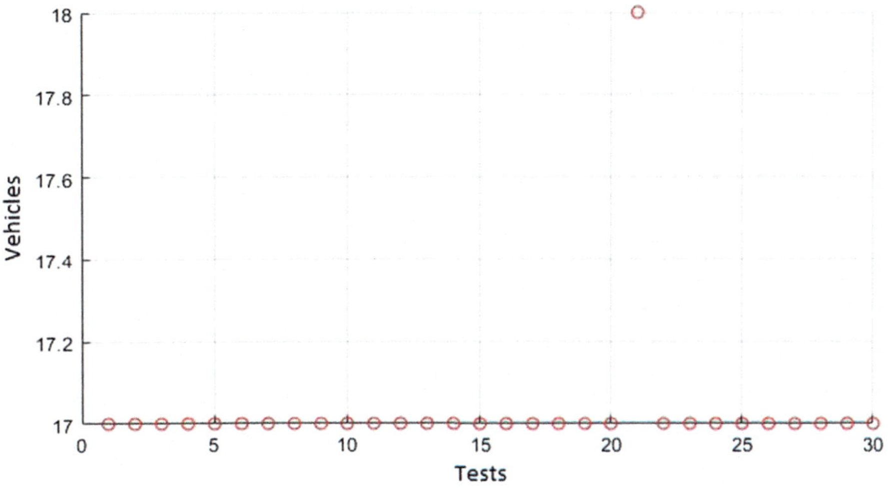

Fig. 2 Behavior of the number of vehicles for genetic algorithm

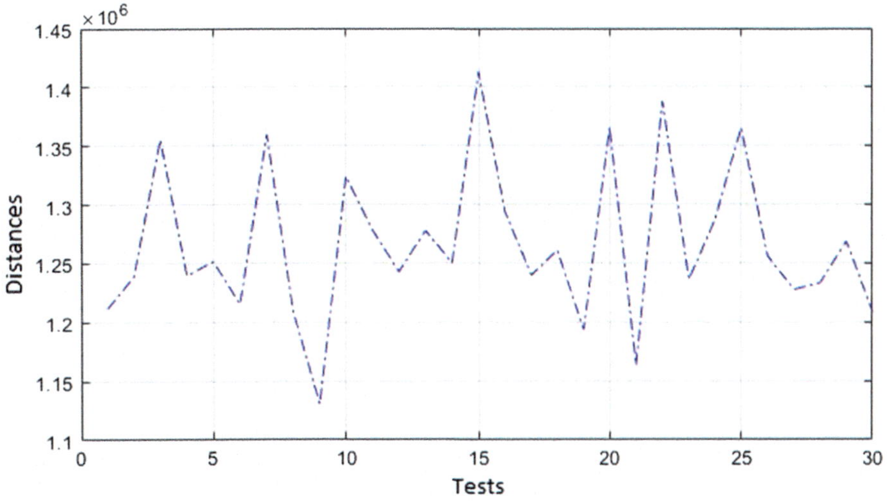

Fig. 3 Behavior of the total distances for genetic algorithm

In Figs. 4 and 5, we show the behavior of the objective function (number of vehicles and total distance) in the executions of 30 runs for the MOGA.

In Table 5, we present the characteristics of the current routes obtained by the GA, the Euclidian total distance traveled for each route and the numbers of workers per route. In average each route traveled 6,648.68 m. The route that traveled the largest distance is the route 13 with 9,706.7 m, and the route with the shortest traveled distance is the route 16 with 3,731.8 m. The total distance traveled for all current routes is of 339,223.91 m.

Table 4 Results of the number of vehicles, total distance and execution time for the multi-objective genetic algorithm

Test	Vehicles	Total distance	Execution (h)
1	17	119,500	6.71
	18	110,260	
	19	109,630	
	20	108,770	
2	17	114,690	5.21
	18	112,240	
3	17	113,880	6.28
	18	112,790	
	19	112,690	
	20	112,460	
4	17	105,080	10.61
	18	104,440	
	19	104,250	
5	18	105,200	5.41
	19	104,480	
6	17	109,140	5.35
7	17	107,280	10.48
	18	106,720	
	19	106,450	
8	18	113,410	5.34
	19	113,330	
9	17	120,080	6.08
	18	114,640	
	19	114,330	
10	17	117,400	5.3
	18	112,730	
	19	112,530	
	20	112,520	

(continued)

In Table 6, we present the characteristics of the current routes obtained with MOGA, the Euclidian total distance traveled by each route and the number of workers per route. In average each route traveled 6,172.51 m. The route that traveled the largest distance is the route 9 with 8,542.8 m, and the route with the shortest traveled distance is the route 14 with 2,309.3 m.

Table 4 (continued)

Test	Vehicles	Total distance	Execution (h)
	21	112,470	
11	18	110,280	5.21
12	17	104,930	5.15
	18	101,660	
13	17	114,710	5.31
	18	112,630	
	19	110,090	
14	17	119,930	5.25
	18	104,480	
	19	102,570	
15	17	111,080	6.39
	18	107,170	
	19	106,540	
16	17	117,400	8.81
	18	112,730	
	19	112,530	
	20	112,520	
	21	112,470	
17	18	110,280	9.06
18	17	104,930	9.18
	18	101,660	
19	17	114,710	10.56
	18	112,630	
	19	110,090	
20	17	117,400	10.16
	18	112,730	
	19	112,530	
	20	112,520	
	21	112,470	
21	18	110,280	8.51
22	17	104,930	8.59
	18	101,660	
23	18	110,120	5.21
24	17	125,210	6.29
	18	111,450	

(continued)

Table 4 (continued)

Test	Vehicles	Total distance	Execution (h)
	19	110,610	
25	17	114,240	10.61
	18	110,080	
	19	109,100	
	20	108,880	
26	17	107,870	9.7
	18	106,780	
27	18	118,800	5.31
	19	118,610	
28	18	110,640	5.34
	19	109,500	
29	17	111,620	5.33
	18	107,260	
30	18	105,740	7.55
	19	105,640	
	20	105,480	
Min	17	104,930	5.15
Max	21	112,470	10.16
Avg	18.27	110677.9747	7.149

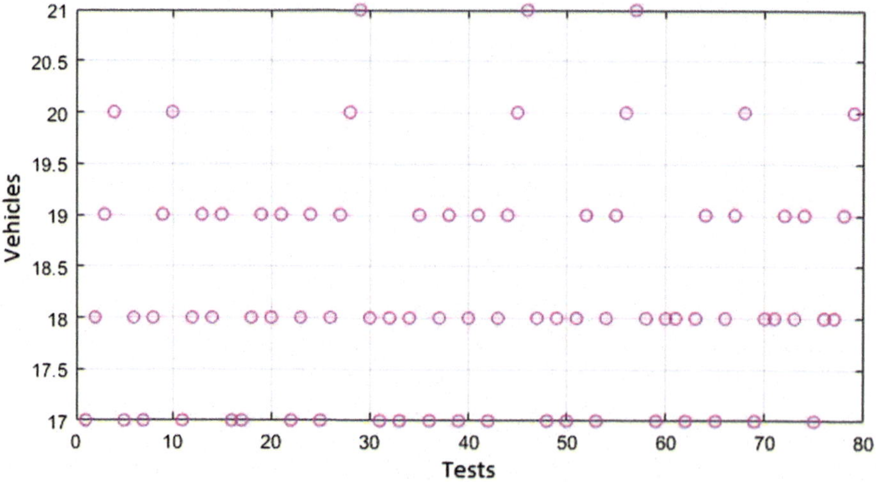

Fig. 4 Behavior of the number of vehicles for the multi-objective genetic algorithm

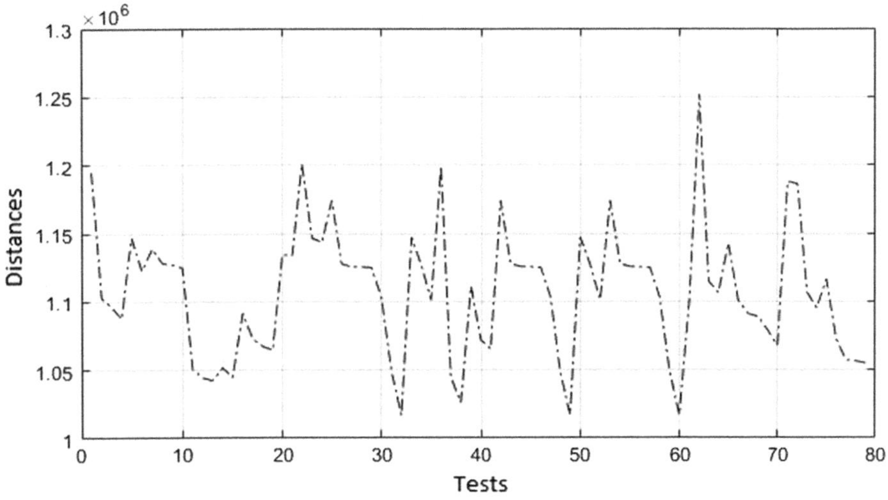

Fig. 5 Behavior of the total distances for multi-objective genetic algorithm

Table 5 Comparison of the traveled distance and workers collected for each route between the real data and the data obtained with genetic algorithm

Vehicle	Current distance	Current workers collected	New distance	New workers collected
1	7,402.22	25	6464	40
2	17,494.28	21	7093.4	40
3	9,911.79	20	6737.8	40
4	3,978.58	31	10169	40
5	14,896.85	29	5467.2	40
6	15,527.28	28	4783.1	40
7	13,534.48	19	6270.2	40
8	7,946.56	19	6112.7	40
9	20,608.94	28	8191.5	40
10	19,829.40	30	5973.4	40
11	11,996.41	25	6719.5	40
12	18,519.64	17	7525.6	40
13	20,184.77	33	9706.7	40
14	18,562.61	32	7026.6	40
15	17161.63	28	6079.3	34
16	13232.11	19	3731.8	29
17	9,691.14	20	4975.8	20
18	17,327.92	19	NA	NA

(continued)

Table 5 (continued)

Vehicle	Current distance	Current workers collected	New distance	New workers collected
19	5,554.54	39	NA	NA
20	15,448.48	12	NA	NA
21	6,346.17	21	NA	NA
22	11,911.61	24	NA	NA
23	8,023.17	13	NA	NA
24	6,622.15	18	NA	NA
25	19,139.02	27	NA	NA
26	2,435.46	21	NA	NA
27	5,936.70	24	NA	NA

Table 6 Comparison of the traveled distance and workers collected for each route between the real data and the data obtained with multi-objective genetic algorithm

Vehicle	Current distance	Current workers collected	New distance	New workers collected
1	7,402.22	25	5294.6	40
2	17,494.28	21	4974.8	40
3	9,911.79	20	6010.4	40
4	3,978.58	31	5776.7	40
5	14,896.85	29	4890.3	40
6	15,527.28	28	5721.4	40
7	13,534.48	19	2309.3	40
8	7,946.56	19	7990.9	40
9	20,608.94	28	8542.8	40
10	19,829.40	30	6180	40
11	11,996.41	25	6644.4	40
12	18,519.64	17	5354.5	40
13	20,184.77	33	7577.9	40
14	18,562.61	32	12135	40
15	17161.63	28	7064.3	36
16	13232.11	19	6150.4	25
17	9,691.14	20	2315	22
18	17,327.92	19	NA	NA

(continued)

Table 6 (continued)

Vehicle	Current distance	Current workers collected	New distance	New workers collected
19	5,554.54	39	NA	NA
20	15,448.48	12	NA	NA
21	6,346.17	21	NA	NA
22	11,911.61	24	NA	NA
23	8,023.17	13	NA	NA
24	6,622.15	18	NA	NA
25	19,139.02	27	NA	NA
26	2,435.46	21	NA	NA
27	5,936.70	24	NA	NA

5 Conclusions

In this paper, we proposed two optimization methods for the Vehicle Routing Problem (VRP), the use of a genetic algorithm and multi-objective genetic algorithm applied to optimize the routes of collected workers to transport them to the plant JABIL Circuit.

The number of vehicles of 17 and 17, and a distance traveled of 113,030 and 104,930 of the genetic algorithm and multi-objective genetic algorithm, respectively, are better than the results of the real data of 27 vehicles and 339,223.91 of total distance.

Genetic algorithms are a strong tool for search procedures. Their performance in terms of execution time is very good since it produces solutions in acceptable times. In the case of GA, the results are more conservative but their execution time is less than half compared to GA Multi-objective. Regarding GA Multi-objective, the solutions surpass those of GA in terms of travel when the number of vehicles is the same. Independently, the behavior of both algorithms is satisfactory and must be taken into consideration when solving similar problems to VRP.

Appendix

```
*Parametres
*[] chromActual: vector whith vehicles data and points
*integer CountP: total of points (include depot)
*integer CountV: max amount of vehicles posible
*integer VC: maximum capacity of vehicles (constraint)
*[] D: vector of demands in each points
*integer TD: total demand to pick up (~sum(D))
*[][] MDist: matrix of distances between all points
begin FO (chromActual, CountP, CountV, VC, D, TD, MDist)
 start = 1;
 countV = 0;
 demandT = 0;
 distanceT = 0;
 //depot or start point.
 depot = 1;
 i = 0;
 visitedPoints = [];
 while (i < CountV) do
   /*vehicle between start and (start+CountP-1)*/
   vehicle = chromActual(start : start+CountP-1);
   demand = 0; //accumulated demand in vehicle
   distance = 0; //accumulated distance in vehicle route
   lastPoint = depot;
   if vehicle(1) == 1 then //is active?
    j = 2;
    while (j < CountP) do
      point = vehicle(j);
      if ( visitedPoints.notcontains(point)
          and (demand + D(point) <= VC)
          and (isRealPoint(vehicle(j))) then
        demand = demand + D(point);
        distance = distance + MDist(lastPoint, point);
        lastPoint = point;
        visitedPoints << point;
      end if
      j = j + 1;
    end while
    demandT = demandT + demand;
    distanceT = distanceT + distance;
    countV = countV + 1;
   end if
   /*move to next vehicle*/
```

```
    start = start+CountP;
    i = i + 1;
  end while
  if (demandT == TD) then //is feasible?
   /*In GA case, R is length(MaxDistancePosible)*/
   if GA ? then
     fitness = (countV*(10^R))+distanceT;
   else if MultiGA ? then
     fitness = [countV, distanceT];
   end if
   return fitness;
  else
   if GA ? then
     return Inf;
   else if MultiGA ? then
      return [Inf, Inf];
```

References

1. Laporte, G.: The vehicle routing problem: an overview of exact and approximate algorithms. Eur. J. Oper. Res. **59**, 345–358 (1992)
2. Li, L., Fu, Z.: The school bus routing problem: a case study. J. Oper. Res. Soc. **53**, 552–558 (2002)
3. Vaira, G.: Genetic Algorithm For Vehicle Routing Problem. Vilnius University, Technological Sciences, Informatics Engineering **1**(1), 2–94 (2014)
4. Park, J., Kim, B.: The school bus routing problem: a review. Eur. J. Oper. Res. **202**, 311–319 (2010)
5. Raghavendra, A., Krishnakumar, T.S., Muralidhar, R., Sarvanan, D.: A practical heuristic for a large scale vehicle routing problem. Eur. J. Oper. Res. **57**, 32–38 (1992)
6. Schittekat, P., Kinable, J., Sorense, K., Sevauz, M., Spieksma, F., Springael, J.: A metaheuristic for the school bus routing problem with bus stop selection. Eur. J. Oper. Res. **229**(2), 518–528 (2013)
7. Baker, B.M., Ayechew, M.A.: A genetic algorithm for the vehicle routing problem. Comput. Oper. Res. **30**(5), 787–800 (2003)
8. Zuñiga-Aguirre, E., González-Gurrola, L.C.: Un algoritmo GRASP para un problema de rutas de vehículos escolares aplicado al transporte de personal de una empresa de manufactura. Recent Advances in Theory, Methods and Practice of Operations Research (2014)
9. Veldhuizen, D.V., Lamont, G.B.: Multiobjective evolutionary algorithms: analyzing the state-of-the-art. Evol. Comput. J. **8**, 125–148 (2000)
10. Saravanan, R., Asokan, P., Sachidanandam, M.: A multi-objective genetic algorithm (GA) approach for optimization of surface grinding operations. Int. J. Mach. Tools Manuf. **42**(12), 1327–1334 (2002)
11. Bektaş, T., Elmastaş, S.: Solving school bus routing problems through integer programming. J. Oper. Res. Soc. **58**, 1599–1604 (2007)
12. Cordeau, J.F., Laporte, G., Savelsbergh, M.W.P., Daniele, V.: Vehicle routing. In: Handbook in Operation Research and Management Science, Elsevier, vol. 14, pp. 367–428 (2007)
13. Dantzig, G.G., Ramser, J.H.: The truck dispatching problem. Manag. Sci. **6**(1), 80–91 (1959)

14. Fisher, M.L.: Optimal solution of vehicle routing problems using minimum k-trees. Oper. Res. **42**(4), 626–642 (1994)
15. Garey, M.R., Johnson, D.S.: Computers and Intractability: A Guide to the Theory of NP Completeness. Series of Books in the Mathematical Sciences, p. 340. Macmillan, New York (1979)
16. Wren, A., Carr, J.D.: Computers in Transport Planning and Operation. Ian Allan, London, UK, pp. 152 (1971)
17. Goldberg, D.E.: Genetic Algorithm in Search Optimization and Machine Learning. Addison – Wesley, New York (1989)
18. Han, S., Tabata, Y.: A hybrid genetic algorithm for the vehicle routing problem with controlling lethal gene. Asia Pac. Manag. Rev. **7**(3), 405–426 (2002)
19. Lozano, J., González-Gurrola, L.C., Rodríguez, E., Lacomme, P.: A statistical comparison of evaluation functions for the vehicle routing problem with route balancing. In: 15th Mexican International Conference on Artificial Intelligence, MICAI (2016)
20. Binitha, S., Sathya, S.S.: A survey of bio inspired optimization algorithms. Int. J. Soft Comput. Eng. (IJSCE) **2**(2), 137–151 (2012)
21. Coello-Coello C.A., Lamont G.B., Veldhuizen D.V.: Evolutionary Algorithms for Solving Multi-Objective Problems, p. 800. Springer, USA (2007)
22. Conak, A., Coit, D., Smith, A.: Multi-objective optimization using genetic algorithms: a tutorial. Reliab. Eng. Syst. Saf. **91**(9), 992–1007 (2006)
23. Deb, K., Deb, K.: Multi-objective optimization. In: Burke, E., Kendall, G. (eds.) Search Methodologies, pp. 403–449. Springer, Boston, MA (2014)
24. Amouzgar, K.: Multi-objective optimization using genetic algorithms. M.S. thesis, Jönköping University, School of Engineering, Jönköping, Sweden (2012)

Agent-Based Model of Resistant Bacterial Evolution in an Heterogeneous Medium

Rubén A. Castañeda-Martínez, Dora-Luz Flores, Carlos Castro and Balam Benítez

Abstract In this chapter, an agent-based model was developed using NetLogo to study bacterial evolution and the process in which bacterial populations are able to acquire antibiotic resistance in an heterogeneous medium with an escalated antibiotic gradient. Molecular interactions between wild-type (WT) *E. coli* and the antibiotic trimethoprim (TMP) were abstracted and conceptually represented to correctly model the complex biological system and interactions. The antibiotic gradient was spatially separated in a plate, starting from $1 \times$ minimal inhibitory concentration (MIC) of WT *E. coli* and scaling by 10-fold up to $1000 \times$ MIC. A set of experiments was performed under the same initial conditions, varying between five different *dnaQ* alleles, to measure the mean antibiotic resistance, bacterial population, deaths caused by antibiotic and deaths caused by the lack of nutrients to determine if the simulations had biological meaning. One hundred simulations were carried out for every allele. Each run of the model had different, but similar results, meaning that the inherent variability was eliminated and the overall behavior of the model was properly characterized. Tukey tests were performed to measure significant differences between results from the different alleles. A typical bacterial growth curve with well defined phases was obtained. Bacteria were able to acquire antibiotic resistance and migrate from the borders towards the center of the plate, where the concentration of TMP was $1000 \times$ MIC. Those results alongside the graphical environment gives enough prove that the model is well implemented to study the bacterial evolution and the process in which they are able to develop antibiotic resistance.

R. A. Castañeda-Martínez
Ensenada, Mexico
e-mail: rubenc1294@gmail.com

D.-L. Flores (✉) · C. Castro · B. Benítez
Autonomous University of Baja California, Ensenada, Mexico
e-mail: dflores@uabc.edu.mx

C. Castro
e-mail: carlos.castro98@uabc.edu.mx

B. Benítez
e-mail: angel.benitez@uabc.edu.mx

M. A. Sanchez et al. (eds.), *Applied Decision-Making*, Studies in Systems, Decision and Control 209, https://doi.org/10.1007/978-3-030-17985-4_3

1 Introduction

Antibiotics have played a relevant role in human health since its discovery 70 years ago. They can be used as a method for treatment and prevention of a great variety of infections, including those that are a mortal threat to humans. It is clear that antibiotics have contributed positively in the life expectancy of the world population [3]. However, in recent years there has been a growing concern about the emergence of different bacterial strains with antibiotic resistance (AR), a situation that has become a public health problem [15, 16].

1.1 Antibiotic Resistance Development

It is necessary to understand how and where bacterial strains with AR are originated and transmitted between different locations. Bad sanitation practices in hospitals and communities, alongside horizontal gene transfer contribute to the propagation of bacteria with AR [19, 24]. Other causes may include auto-medication, improper use of antibiotics and high costs to acquire good-quality medical services. It is not clear how bacteria propagate and evolve so quickly to survive to antibiotics [16]. Understanding the underlying mechanisms of resistant strains evolution and propagation could allow the development of new, safe, and effective generations of antibiotics. These antibiotics could have a better span life in the clinical area, preventing the evolution and propagation of strains with AR [24].

1.2 Heterogeneous Environments

On another topic, heterogeneous environments are spatial environments with different pharmacokinetic parameters that facilitate the evolution of AR in bacteria [26]. They can be found in human organs and tissues. The same goes for patients, due to the unique immune response of each individual [16]. Due to he high compartmentalization of the human body, we can find different environments with different levels of isolation, leading to a group of variable stress gradient environments, where environments with lower stress can foster a mode of adaptation.

These environments could propitiate a fast propagation of bacteria between patients in hospitals [19]. According to Tung et al. [26], there are two main characteristics in these environments. The first one is a stress gradient in the environment, which allows mutant strains to acquire resistance and then develop relative fitness when moving to an environment with higher stress level, leading to the second characteristic; because of the relative small size of higher stress environments, it is easier for mutants to rapidly adapt to these environments.

Two important variables to take into account when talking about heterogeneous environments are the adaptation and range. The adaptation to new stress levels enable mutants to enlarge their range by evading competition when invading previously inaccessible habitats and providing with new features to the mutant in instance. This adaptation and range are enhanced by the heterogeneity of the environment providing a staircase to resistance [16].

In Baym et al. [5], a novel plate to study migration and evolution of *Escherichia coli* in heterogeneous environments is presented. The microbial evolution and growth arena (MEGA) plate consists of an antibiotic (ciprofloxacin or trimethoprim) concentration gradient that increases from the right and left edge to the center of the plate. This device allows the study of spatiotemporal evolution of bacterial strains and the development of AR in heterogeneous environments.

1.3 Development of Antibiotic Resistance at the Molecular Level

Trimethoprim (TMP) is a bacteriostatic antibiotic that has a high affinity to the bacterial enzyme dihydrofolate reductase (DHFR). Its mechanism of action consists of selectively inhibiting DHFR by competing with his substrate, dihydrofolate. This enzyme catalyzes the reduction of dihydrofolate to tetrahydrofolate with NADPH, and is part of the biosynthesis of purines, pyrimidines and some amino acids [25]. By blocking the activity of DHFR, the biosynthesis of purines is reduced and eventually the DNA replication and cell division will be stopped, leading to cell death.

Several mechanisms of resistance acquisition to TMP have been identified. The most commons are: reduction of the cell permeability, reducing the amount of TMP that enters into the cell; activating efflux pumps to remove the antibiotic from the cell; overexpression of DHFR; and the phenomenon that is the research objective of this work, the expression of DHFR with an altered conformation and therefore a lower affinity to TMP. This phenomenon occurs by two mechanisms: mutations in the *folA* gene, which codifies DHFR, and TMP-resistant *folA* gene acquisition by horizontal transfer.

The presence of TMP activates the most common resistance mechanisms which are the mutations in the *dfr* gene family, responsible for coding the DHFR [11]. SOS response system is related to mutagenesis due to the DNA repair system which is activated when the bacteria is exposed to stress. When the SOS system is triggered, bacteria produce defense proteins (whose genes are normally in a repressed state) allowing the repair of damaged DNA and reactivation of DNA synthesis. These SOS genes expression increases when either the chromosomal DNA is damaged or when the DNA replication is blocked, but also it has been demonstrated that SOS system can be induced when the bacteria is starving, leading to adaptive mutants. During starvation, signals for SOS system are created but to induce the system, the cells must be in a dividing phase and under conditions that allows the generation of high concentrations of cAMP, meaning that there is not enough carbon source in the medium [17, 22].

In their study, Baym et al. [5] found that the most frequently mutated gene in the resistant strains was *folA*. The expressed mutated DHFR has a lower affinity for TMP, meaning that less molecules of TMP can bind to DHFR and thus the bacteria acquires resistance to TMP. Also, a positive correlation was found between the number of mutations (SNPs, single-nucleotide polymorphisms) in the *folA* gene and the quantity of TMP that the bacterium can resist [5]. Another interesting finding of Baym et al. [5] is that all highly mutated isolates of *E. coli*, but none of the others, had a mutated allele of the *dnaQ* gene, that codifies for the ε subunit of the DNA polymerase III, which has exonuclease activity and is responsible of proofreading in the DNA replication. It has been previously reported that expression of mutated *dnaQ* alleles triggers the DNA repair SOS induction system, as well as increasing mutation frequency by up to 1600-fold times as high compared with the wild-type (WT) strain [12].

When a missense mutant ε protein is expressed, the proofreading function can be reduced or completely lost, thus the DNA polymerase is unable to correct errors during the DNA replication, which causes a significant increase in mutagenesis [12]. This phenomenon is correlated with the acquisition of antibiotic resistance; when the DNA is replicating, the ε subunit is unable to remove incorrectly incorporated bases in the *folA* gene which are fixed in the generations, and mutated DHFR proteins are expressed. Over generations, eventually a TMP-resistant DHFR protein will express, allowing the emergence of bacterial strains with AR capable to migrate to the next section of the plate.

According to the study of Gautam et al. [12], the SOS induction leads to the expression of a transposase whose activity allows *E. coli* cells to acquire resistance to chloramphenicol, as well as increasing mutagenesis. As stated by their conclusions, the increase in mutation frequency is probably due to the DNA polymerase III defective proofreading activity. Different mutant alleles increase mutagenesis in different ratios. To run the simulations, we took into account D12G, D103G, V96G, and T151 *dnaQ* alleles, which increase mutation frequency in 1695, 1108, 20 and 312 times the WT mutation frequency, respectively [12]. This point is better explained in the methodology section.

1.4 Agent-Based Modeling

A model is defined as an abstract and simplified representation of a given reality. The type and complexity of a model directly depends on the question you want to answer [1]. Computational models have the advantage to solve mathematical models and be superior to them because the abstract representation achieved in coding sometimes could not be described or represented in other way [14]. Computer models are an alternative to study biological complex systems substituting mathematical models made with ordinary differential equations (ODE). In addition, computer models can support in vivo and in vitro experimentation by reducing the number of experiments made [2]. On the other hand, a simulation is the perceptible manifestation of a model, represented by a computer program that provides ideas about a system [1]. Within the in silico models there are two types: the deterministic ones, which produce identical output profiles with identical status and initial conditions; and stochastics, which allow random or probabilistic events, with which a family of response profiles can be obtained in a single simulation with an identical initial state and conditions.

When implementing biological systems models, several factors must be considered that will define the complexity and output (or outputs) of the model, such as the question that is wanted to answer, the problem context, the type of response you want to obtain and the available experimental data. If the model is very simple, it may not be possible to represent the underlying relationships between the various variables of the phenomenon that the data provides. At the other extreme, if it is very complex or detailed, the data could not be compatible with the structure of the model or the estimation of the output parameters, which would generate a perfect fit to the data and with this the inability to respond faithfully to new data not previously shown, and a model that is unnecessarily difficult to solve in terms of computational processing [29]. Given the nature randomization that characterizes natural phenomena, stochastic models are generally most appropriate for modeling biological phenomena, considering the inherent noise and heterogeneity in these systems.

Agent-based modeling (ABM) is a type of stochastic computer model, where autonomous agents take decisions based on rules to accomplish a specific behavior of a real-life phenomenon. This method allows the spatiotemporal analysis of phenomenon. In ABM, analysis can be done integrating qualitative information and quantitative data [8]. ABM has been used to model biological systems in different areas, like brain tumors [30], cell cycle [8], muscle contraction [18], epidemic evolution [7], host-pathogen systems [4] and cell culture bioreactors [6] among others.

In this work, we used ABM to build a computational model in NetLogo to study bacterial spatiotemporal evolution and antibiotic resistance acquisition in an spatially structured heterogeneous environment, following the characteristics of the MEGA plate [5].

2 Methodology

Simulation as a research method is a general methodology where real-life data is compared against the results from the simulation of a model of a given phenomenon. In order to validate the model, a similarity between collected data and the simulation results is expected [13]. ABM of the microbial evolution was created following the methodology from Wilensky and Rand [28].

The environment of the model represents an agar medium, segmented into nine blocks of the same width. The first and last segments lack TMP. Starting from each side of the plate, the concentration magnitude of TMP starts at $1 \times$ minimum inhibitory concentration (MIC) of WT *E. coli*, and then it increases 10-fold for the next segment until having $1000 \times$ MIC.

In NetLogo, the world consists of a grid of squares of the same size, each patch can have any desired properties that fit the phenomenon that is going to be modeled. In this model each patch have the same initial concentration of nutrients. Time is measured in ticks, that represent the execution of one code cycle that varies depending on the computer processor running the simulation.

Agents represent the bacteria in the medium. Each bacterium has the following properties:

1. Affinity: Affinity of the enzyme dihydrofolate reductase for the sustrate.
2. State: Determines if a bacterium is alive (white) or dead (yellow).
3. Energy: Amount of energy that bacteria have for movement.
4. Age: Age of a bacterium measured in ticks.

Bacteria have a chemotaxis-based movement [5, 21], moving to a patch that has a higher amount of nutrients than the patch on which it is currently standing. When the amount of nutrients in the patches around a bacterium are the same, his movement is a Brownian (stochastic) movement [23].

Bacteria take nutrients when they are in a patch to collect energy. They execute a reproduction process when they have the necessary energy to do it. This event is conditioned to a probability called reproduction rate.

Bacteria can develop AR through mutations that lower the value of the affinity property; a low level of affinity will make the bacteria able to survive in higher amounts of antibiotic. Negative mutations are also possible, making a bacterium to lose its AR. This process generates a selection of resistant strains, in which the mutations that granted AR are fixed in other generations of bacteria.

When a bacterium stands on a patch that has a higher amount of TMP that the one that she can survive, it dies. When this happens, the concentration of TMP decreases a small amount, to represent the inactivation of TMP molecules when they bind to DHFR. Bacteria can also die from starvation, when their energy is depleted and there are not enough nutrients in the environment.

Programming of the processes defined in the conceptual model was performed with a specific commands and functions in order to accurately reproduce the real phenomenon. The model was built in NetLogo version 6.0.4 [27].

The model consists of four inputs previously described, which are shown in Table 1, altogether with the value used in the simulation and when applies, the reference from where that value was obtained. The initial inoculum represents the initial number of bacteria at the beginning of the simulation. The mutation rate is the probability of each bacterium to mutate, which evaluates in every tick. The reproduction rate is the probability that each bacterium do cell division. The values of initial inoculum and reproduction rate were chosen after model calibration and for the sake of the model's simplicity. A higher initial inoculum and reproduction rate would cause more cell divisions and considerably more agents, causing the simulations to slow down and taking more time to finish. Lower initial values would mean not enough bacteria to reach the next region and the death of the population. This phenomenon is known as chaos, which means a system is very sensible to small changes in the initial conditions, a typical characteristic of complex systems [28].

Table 1 Values used for the inputs during the simulation of the model

Input	Value	Reference
Initial inoculum	20 bacteria	
Mutation rate	$1.6 \times 10^{-5}\%$	Lee et al. [20]
Reproduction rate	5%	
Allele	Wild-type (1)	Gautame et al. [12]

Table 2 *dnaQ* alleles studied in the simulations

dnaQ allele	Mutation frequency ratio[a] (*dnaQ*/wild-type)
Wild-type	1
V96G	20
T151	312
D103G	1108
D12G	1695

[a]Mutation frequency ratio is the mutation frequency of strains with *dnaQ* alleles in comparison with mutation frequency from wild-type strain. Data was taken from Gautam et al. [12]

The *dnaQ* allele represents a multiplier for the mutation rate by a specific number for each allele. When the simulation is run with an allele different than the WT, bacteria have two different mutation rates: the WT rate, when there is no antibiotic present, and the one that the allele confers, to represent the activation of the SOS induction in presence of antibiotic.

Inputs in the present model are set as free parameters to allow the user to choose the initial values if another specific bacterium wants to be studied. In addition, free parameters are an option to test a specific phenomenon with values that are impossible or difficult in real life, in order to predict behaviors in different initial values.

On the other hand, the outputs of the model consists of the image of the growth medium, that can be used to see the bacterial distribution and reproduction in the medium; and four graphs, that shows the mean antibiotic resistance, the total bacteria population, deaths caused by the antibiotic and deaths caused by starvation.

An experiment was performed by running the model 100 times with the initial values from Table 1 for each allele (WT, V96G, T151, D103G, and D12G). Mutation frequency ratio of *dnaQ* alleles are shown on Table 2.

3 Results

The appearance of bacterial lineages with AR was observed when running the model with inputs from Table 1. Newer generations of bacteria were able to survive in the sections with the highest antibiotic concentration, i.e. 1000 × MIC of TMP. Figure 1 shows the world of the model after completing a simulation. On the right side, the population was unable to mutate in a rate fast enough to allow them to

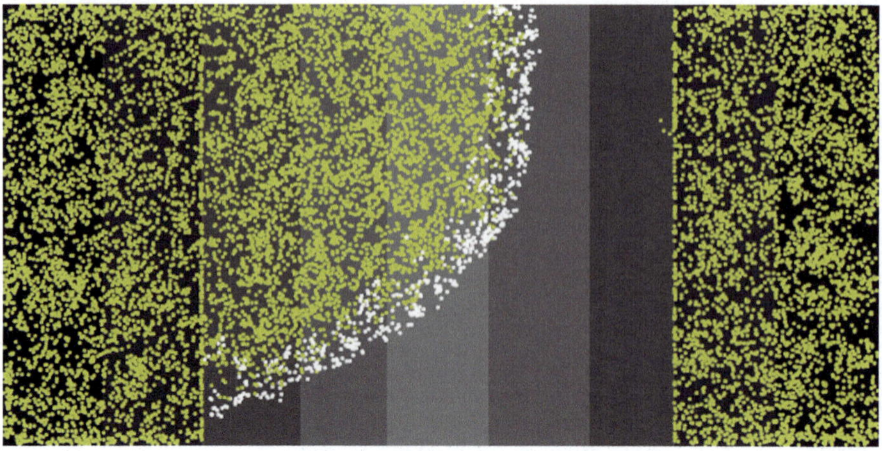

Fig. 1 Bacterial population dynamics in the antibiotic medium

migrate onto the next region before the complete depletion of the nutrients. On the left side, bacteria mutated faster and were able to reach the central section, which has 1000 × MIC. This phenomenon occurred in some simulations but it is not a constant. In most simulations, specially in the ones when the *dnaQ* alleles were studied, bacteria were able to migrate towards the center of the plate from both ends. Similar results to the findings of Baym et al. [5] were observed.

3.1 Simulations

The different outputs obtained by the model for the WT *dnaQ* allele can be seen in Fig. 2. The mean antibiotic resistance of the microbial population is shown in Fig. 2a, with every line representing one simulation. Mean antibiotic resistance seems to increase over time, a process that is correlated with the development of AR on bacteria occasioned by mutations on the genome. Bacteria acquire AR when they migrate towards regions with higher antibiotic concentrations, where mutations are more likely to occur due to the activation of the SOS induction and mutations in the *dnaQ* gene.

The changes in the bacteria population during the simulation are shown in Fig. 2b. Populations are shown on a logarithmic scale for an easier visualization. A typical bacteria growth curve was obtained: there is a well-defined fast exponential phase, followed by a stationary phase where cell division and death occurs in similar proportions, finishing with a death phase where bacteria die. As it can be seen, the population dies completely at different amount of ticks on different simulations, ranging from 600 ticks to more than 2000, due to the stochastic nature of the model.

The number of deaths in Fig. 2c are caused by the inactivation of the DHFR by TMP. As it can be observed, there is an exponential grow in the beginning of

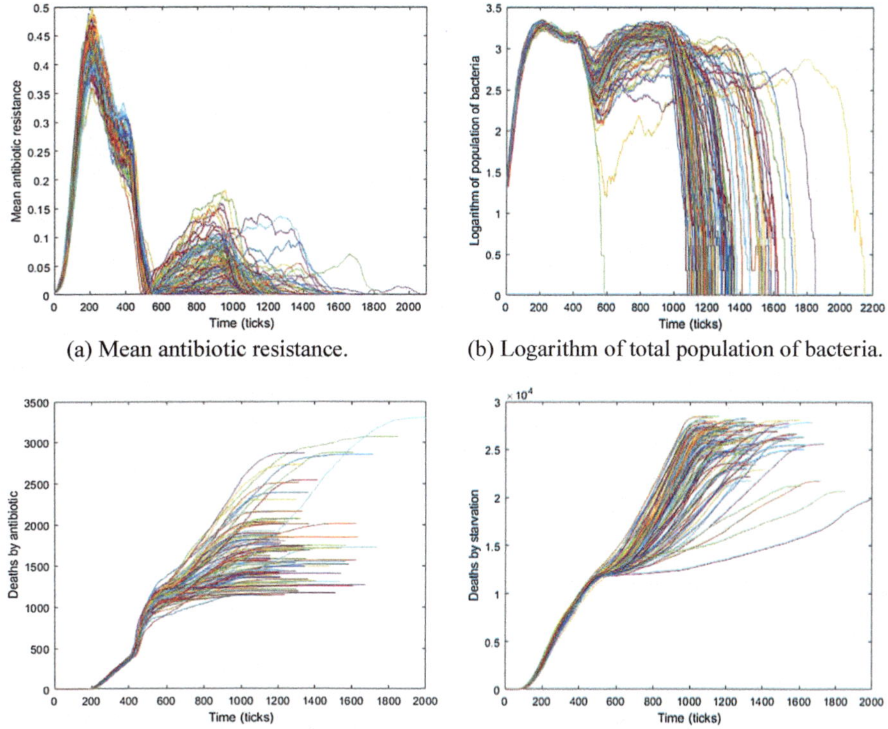

(a) Mean antibiotic resistance.

(b) Logarithm of total population of bacteria.

(c) Number of deaths caused by antibiotic.

(d) Number of deaths caused by starvation.

Fig. 2 Outputs from the model for the wild-type allele. **a** Average antibiotic resistance of the total population over time. **b** Bacterial growth curve, shown in logarithm of the total population. **c** Number of bacteria that have died by the inactivation of dihydrofolate reductase by trimethoprim. **d** Number of bacteria that have died from starvation

the simulation, and then at near 1000 deaths it stabilizes until the amount of deaths caused by antibiotic is irrelevant compared to the amount of total deaths. While in Fig. 2d the amount of deaths are caused by starvation. As it can be observed, this amount is considerably higher than the amount of deaths caused by the antibiotic.

For the allele V96G, Fig. 3 shows the different outputs obtained by the model. The mean antibiotic resistance of the microbial population is shown in Fig. 3a, mean antibiotic resistance seems to increase over time similar to the WT allele.

Changes in the bacteria population during the simulation are shown in Fig. 3b. For this allele, the bacteria died in times ranging from 1000 to 1200 ticks, less time than with the WT allele.

The number of deaths caused by the inactivation of the DHFR by TMP can be seen in Fig. 3c. As it can be observed, there was less variation than the results obtained in the WT. While Fig. 3d shows the amount of deaths caused by starvation with a strong linearity and less variation than the WT.

For the allele T151, the Fig. 4 shows the different outputs obtained by the model. The mean antibiotic resistance of the microbial population is shown in Fig. 4a, mean

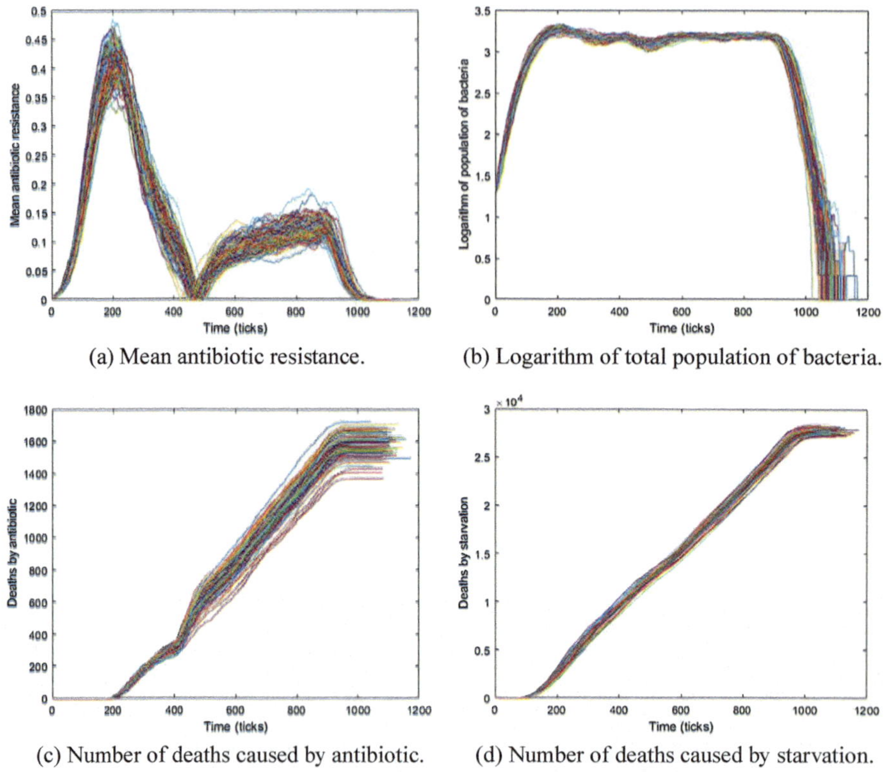

(a) Mean antibiotic resistance.

(b) Logarithm of total population of bacteria.

(c) Number of deaths caused by antibiotic.

(d) Number of deaths caused by starvation.

Fig. 3 Outputs from the model for allele V96G. **a** Average antibiotic resistance of the total population over time. **b** Bacterial growth curve, shown in logarithm of the total population. **c** Number of bacteria that have died by the inactivation of dihydrofolate reductase by trimethoprim. **d** Number of bacteria that have died from starvation

antibiotic resistance seems to increase over time, similar to the results obtained with the WT allele.

The changes in the bacteria population during the simulation can be seen in Fig. 4b. For this allele, the bacteria died in times ranging from 1000 to 1200 ticks, similar to the results obtained from allele V96G.

The number of deaths caused by the inactivation of the DHFR by TMP are shown in Fig. 4c, while Fig. 4d shows the amount of deaths caused by starvation. As it can be seen, there were less deaths caused by starvation than deaths caused by antibiotic, contrary to the results from WT and allele V96G.

The different outputs obtained by the model for the allele D103G are shown in Fig. 5. The mean antibiotic resistance of the microbial population is shown in Fig. 5a, mean antibiotic resistance seems to increase over time, similar to the results from WT allele.

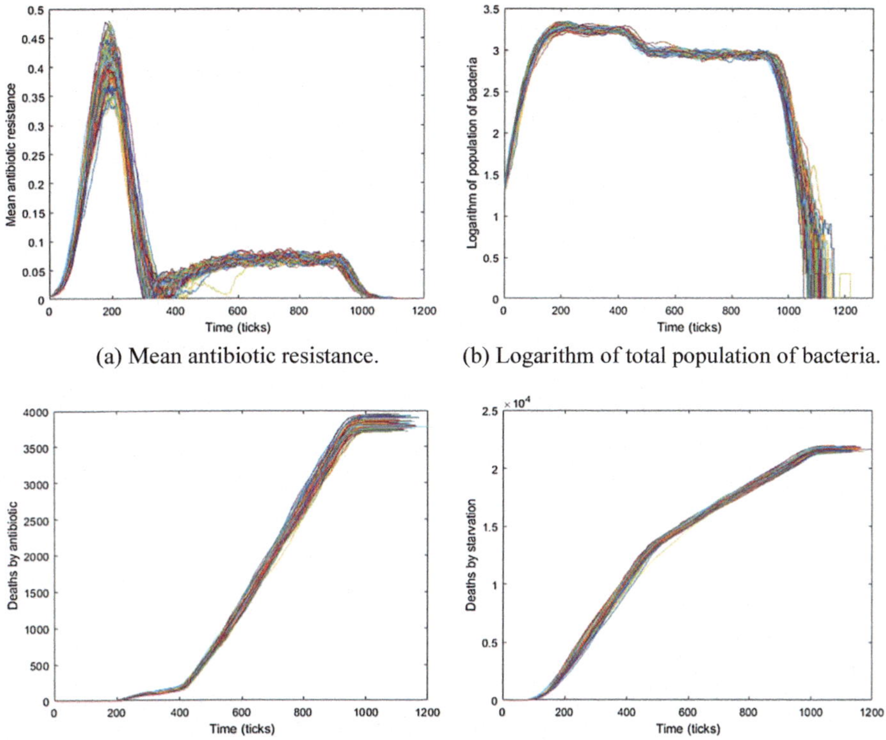

(a) Mean antibiotic resistance. (b) Logarithm of total population of bacteria. (c) Number of deaths caused by antibiotic. (d) Number of deaths caused by starvation.

Fig. 4 Outputs from the model for allele T151. **a** Average antibiotic resistance of the total population over time. **b** Bacterial growth curve, shown in logarithm of the total population. **c** Number of bacteria that have died by the inactivation of dihydrofolate reductase by trimethoprim. **d** Number of bacteria that have died from starvation

The changes shown in Fig. 5b correspond to the bacteria population during the simulation. Whit this allele, the bacteria died in times ranging from 1000 to 1200 ticks, similar times than the allele V96G and T151, but less than with the WT allele.

The number of deaths observed in Fig. 5c are caused by the inactivation of the DHFR by TMP while the amount of deaths caused by starvation are observed in Fig. 5d. As it can be seen, there was less deaths caused by starvation than deaths caused by antibiotics, contrary to the results from the other alleles but similar to the results from allele D12G.

For the allele V96G, Fig. 6 shows the different outputs obtained by the model. The mean antibiotic resistance of the microbial population is shown in Fig. 6a, mean antibiotic resistance seems to increase over time, similar to the WT allele.

The changes in the bacteria population during the simulation can be seen in Fig. 6b. For this allele, the bacteria died in times ranging from 1000 to 1400 ticks, less time than with the WT allele.

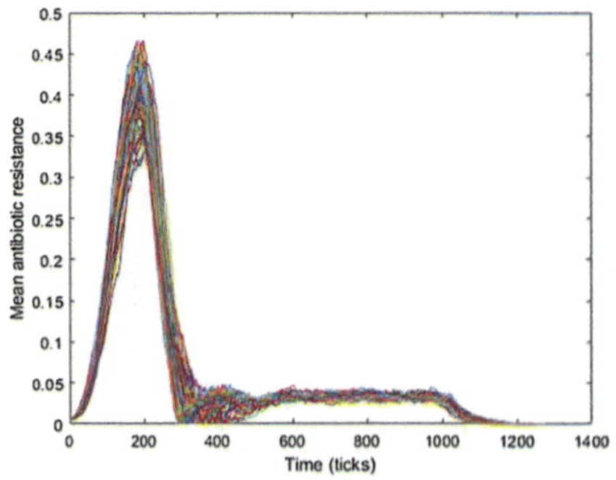

(a) Mean antibiotic resistance.

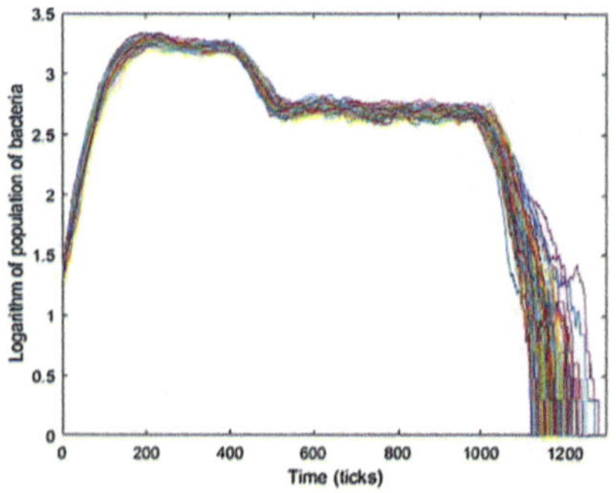

(b) Logarithm of total population of bacteria.

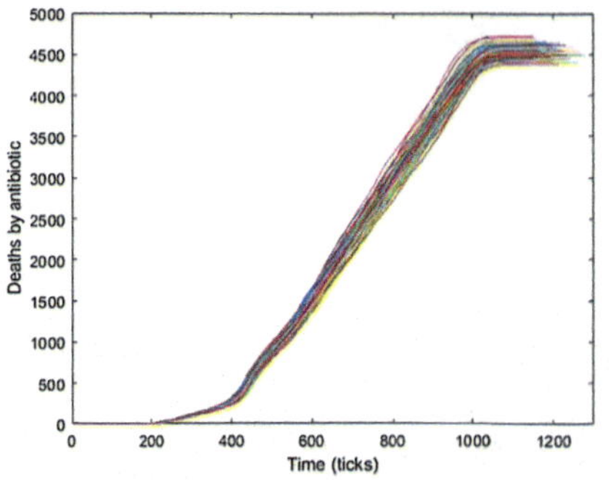

(c) Number of deaths caused by antibiotic.

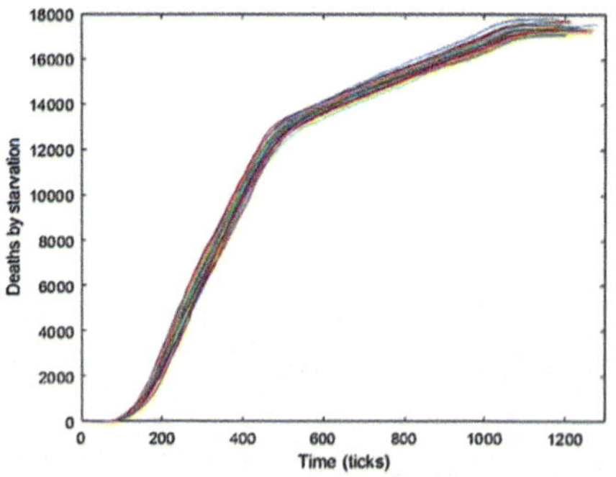

(d) Number of deaths caused by starvation.

Fig. 5 Outputs from the model for allele D103G. **a** Average antibiotic resistance of the total population over time. **b** Bacterial growth curve, shown in logarithm of the total population. **c** Number of bacteria that have died by the inactivation of dihydrofolate reductase by trimethoprim. **d** Number of bacteria that have died from starvation

In Fig. 6c are shown the number of deaths caused by the inactivation of the DHFR by TMP and Fig. 6d shows the amount of deaths caused by starvation. Similar results to the allele D103G.

3.2 *Statistical Analysis*

Some statistical analysis were performed to determine if the different alleles were significantly equal or not. A one-way ANOVA was performed for each of the four output variables to see if the means were statistically equal for the 5 alleles. In addition, a Tukey comparison was performed to compare each allele against the others. The confidence intervals were set to 95%, therefore $\alpha = 0.05$ and the null hypothesis is that all means are equal. For the of deaths caused by starvation, the p-value of the one-way ANOVA was zero, therefore, the means of the five alleles are not equal. In Table 3 are the results for the Tukey comparison for that variable.

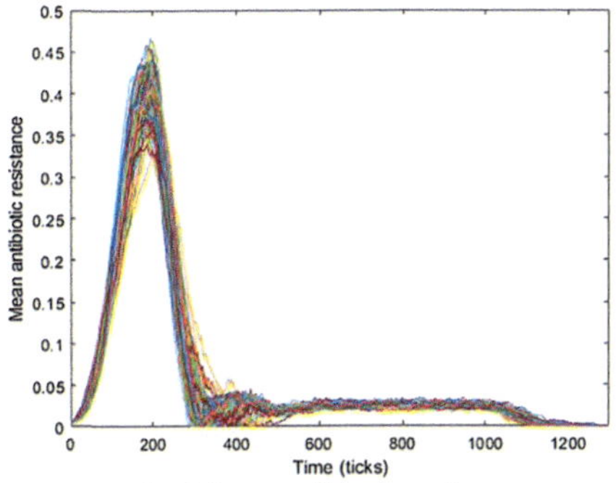

(a) Mean antibiotic resistance.

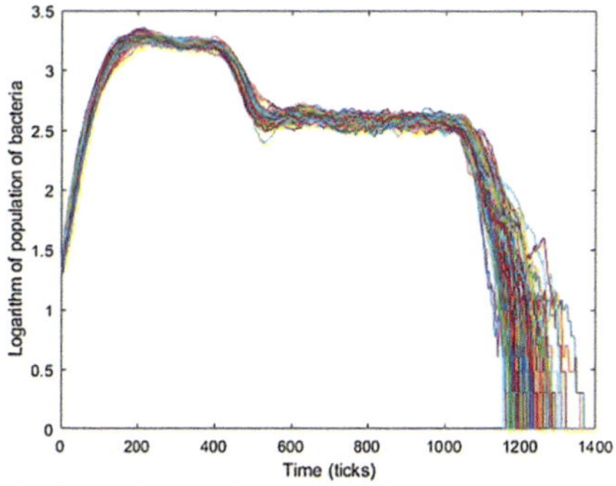

(b) Logarithm of total population of bacteria.

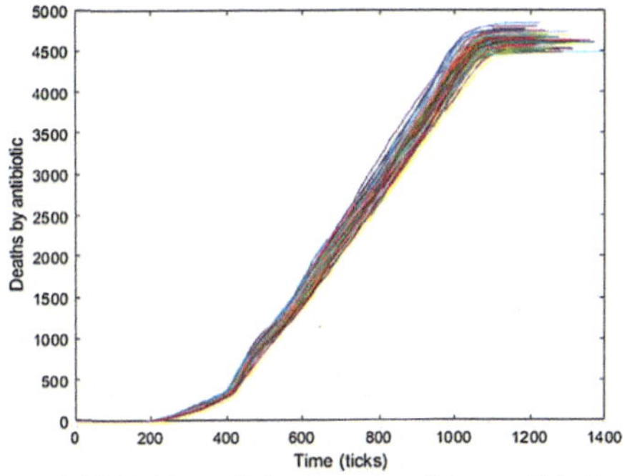

(c) Number of deaths caused by antibiotic.

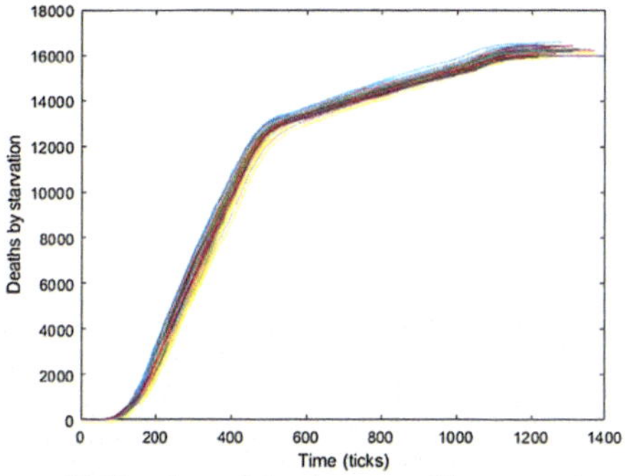

(d) Number of deaths caused by starvation.

Fig. 6 Outputs from the model for allele D12G. **a** Average antibiotic resistance of the total population over time. **b** Bacterial growth curve, shown in logarithm of the total population. **c** Number of bacteria that have died by the inactivation of dihydrofolate reductase by trimethoprim. **d** Number of bacteria that have died from starvation

Table 3 Tukey comparison for deaths caused by starvation

Allele 1	Allele 2	*p*-value
Wild	V96G	0.999
Wild	T151	0.003
Wild	D103G	0.000
Wild	D12G	0.000
V96G	T151	0.013
V96G	D103G	0.000
V96G	D12G	0.000
T151	D103G	0.002
T151	D12G	0.000
D103G	D12G	0.659

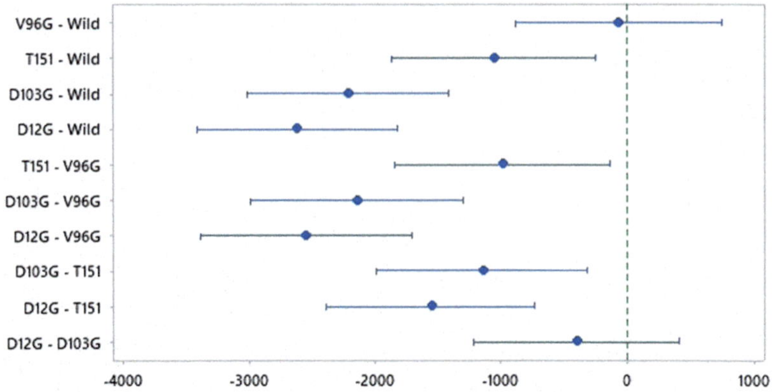

Fig. 7 Tukey test for difference of means of the deaths caused by starvation

Table 4 Tukey comparison for deaths caused by antibiotic

Allele 1	Allele 2	*p*-value
Wild	V96G	0.000
Wild	T151	0.000
Wild	D103G	0.000
Wild	D12G	0.000
V96G	T151	0.000
V96G	D103G	0.000
V96G	D12G	0.000
T151	D103G	0.000
T151	D12G	0.000
D103G	D12G	0.679

For the starvation variable, the WT and allele V96G had a difference of means with $p = 0.999$ so they are statistically similar as well as allele D103G and D12G with a $p = 0.695$. All the other comparisons had a $p < 0.05$, meaning that they are statistically different and therefore the deaths caused by starvation are independent of the mutation frequency ratio for given alleles. All the differences of means can be seen in Fig. 7; where intervals does not contain zero the means are significantly different.

For the of deaths caused by antibiotic the *p*-value of the one-way ANOVA was zero, therefore, the means of the five alleles are not equal. In Table 4 are the results for the Tukey comparison for that variable.

In the alleles from the deaths caused by antibiotics all the comparisons had a $p < 0.05$, except for the allele D103G versus the allele D12G that had a $p = 0.679$, so they are statistically similar. All the differences of means can be seen in Fig. 8; where intervals does not contain zero the means are significantly different.

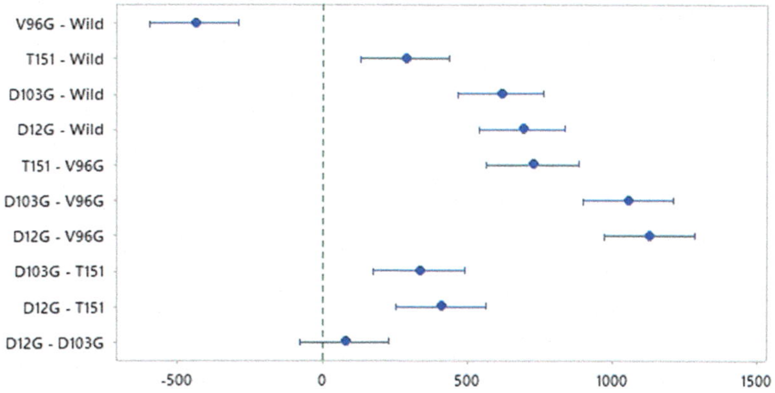

Fig. 8 Tukey test for difference of means of the deaths caused by antibiotic

Table 5 Tukey comparison for the mean antibiotic resistance

Allele 1	Allele 2	*p*-value
Wild	V96G	0.000
Wild	T151	0.019
Wild	D103G	0.096
Wild	D12G	0.002
V96G	T151	0.000
V96G	D103G	0.000
V96G	D12G	0.000
T151	D103G	0.000
T151	D12G	0.000
D103G	D12G	0.771

As for the logarithm of the population, the *p*-value of the one-way ANOVA was zero, the alleles are different. In Table 5 are the results for the Tukey comparison for that variable.

In the population level, the allele D103G and D12G had a $p = 0.771$ as well of allele D103G and WT with a $p = 0.096$ so they are statistically similar between them. In this case, all the other comparisons had a $p < 0.05$, meaning that they are statistically different. All the differences of means can be seen in Fig. 9; where intervals does not contain zero the means are significantly different.

With the mean antibiotic resistance, the *p*-value of the one-way ANOVA was zero, meaning that the alleles are different. Table 6 shows the results for the Tukey comparison for that variable.

For the mean antibiotic resistance, the allele D103G and D12G had a $p = 0.813$ as well of allele T151 and WT with a $p = 0.186$ so they are statistically similar between them. In this case, all the other comparisons had a $p < 0.05$, meaning that they are statistically different. All the differences of means can be seen in Fig. 10; where intervals does not contain zero the means are significantly different.

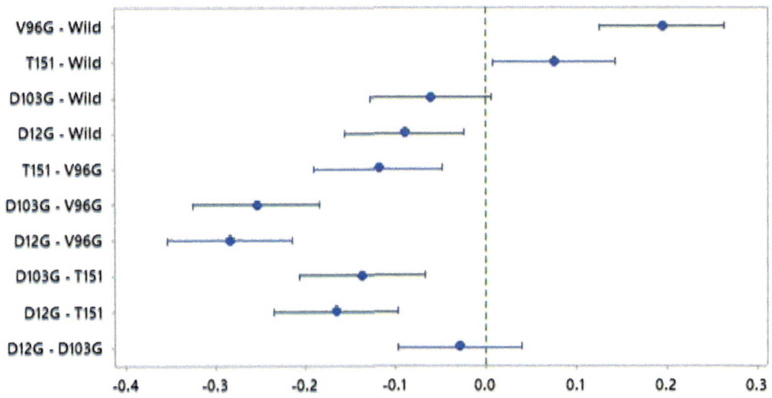

Fig. 9 Tukey test for difference of means of the logarithm of the population

Table 6 Tukey comparison for the mean antibiotic resistance

Allele 1	Allele 2	*p*-value
Wild	V96G	0.000
Wild	T151	0.186
Wild	D103G	0.000
Wild	D12G	0.000
V96G	T151	0.000
V96G	D103G	0.000
V96G	D12G	0.000
T151	D103G	0.000
T151	D12G	0.000
D103G	D12G	0.813

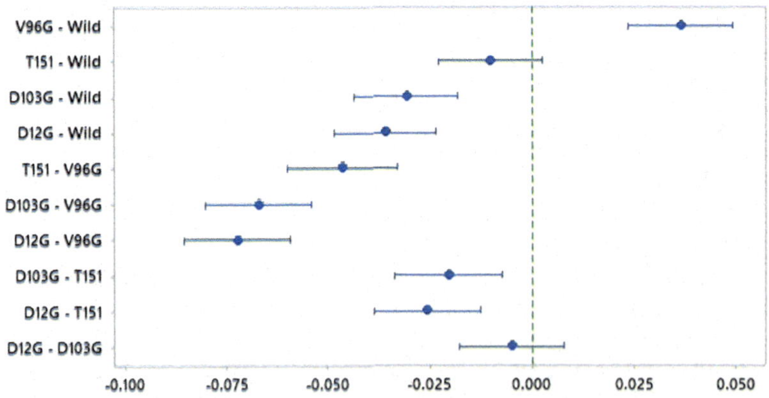

Fig. 10 Tukey test for the difference of means of the mean antibiotic resistance

4 Discussions

One advantage of using ABM is the graphical environment where all the components of the model can be observed, feature not possible in other types of modeling like ODEs. This characteristic allows the model to strengthen one of the attributes of the MEGA plate, i.e. the visual demonstration of bacterial evolution in the antibiotic medium.

Graphs for bacterial population, mean antibiotic resistance and deaths caused by starvation and by antibiotic show a similar behavior in all the runs for each parameter. Even that each run has the same inputs, the results varies due to the model stochasticity. One hundred runs were performed to eliminate the inherent variability of the results in order to properly characterize the biological phenomenon. As it is shown in Fig. 2d, the number of deaths caused by antibiotic starts with an exponential pattern, and then it stabilizes to almost zero deaths at tick 900; this is caused by the development of AR to TMP by bacterial strains.

For all alleles, except the WT, the population died completely in a small range of ticks. In other words, the number of ticks that lasted each simulation was relatively constant in all the alleles except the WT. Possibly it could be that in most simulations of the mutant alleles, the bacterial population managed to migrate to the center of the plate and consume all the nutrients before dying, something that did not happen constantly in the simulations with the WT allele. This may be due to the fact that the low mutation rate of the WT allele does not give the bacteria enough time to acquire resistance to be able to migrate to the next zone. Another interesting event is that bacterial population with the WT allele behaved in a more stochastic way, compared to the other results. As it can be appreciated from the graphs, the simulations with the WT had a higher variation in each of the outputs, compared with the mutant alleles, which had a more defined behavior with a lower variation between simulations.

With respect to the average antibiotic resistance, a drastic fall around 300–400 ticks can be clearly seen in simulations with mutant alleles, which coincides with a slight drop in the total population. This is possibly due to the moment in which the mutant lineages that are left behind consume all the existing nutrients, but they can not continue to advance because they are physically blocked by the mutant lineages that are advancing in front and die from starvation. This phenomenon was clearly characterized in the work of Baym et al. [5], where it was concluded that not necessarily the most adapted or resistant mutants go to the front of the population, but that the first bacteria that manage to migrate are the ones that survive, while that laggards may have a higher resistance to the antibiotic but die due to a lack of nutrients.

When performing the Tukey test, in most cases there was a difference between the results obtained in the model for the different alleles. But there was some cases when the results where not different when having a different allele as initial condition. The most notorious case was with the allele D103G and the allele D12G that had no meaningful differences when performing the Tukey comparisons. This means that the behavior of the bacteria does not change between those two alleles in the case of this model.

The proposed model can be used as complement to studies related to the dynamics and evolution of different types for bacteria in an antibiotic medium. A perfect abstraction of a real phenomenon is usually impossible to achieve, and frequently the most complex processes must be simplified. Some parameters were set as the real biological processes from *E. coli*, but it is possible to change the inputs with values that fit a specific type of bacteria, an advantage of the use of free parameters in the ABM.

5 Conclusions

The ABM of the MEGA plate was created to simulate the dynamics and spatiotemporal evolution of bacterial populations in an medium with heterogeneous antibiotic concentrations. This model was used to study *E. coli* acquisition of AR in a heterogeneous medium, a similar process in which the bacteria mutates in the human body or in a hospital and medical industrial environment. The model allows the study of mechanisms which leads to the development of mutations that endow the bacteria with new qualities and characteristics to survive in high antibiotic concentrations. Understanding this process allows the creation of new strategies to combat the undesired growth of pathogenic bacterial populations and to avoid, if possible, the rapid evolution of resistant strains.

It is clear that environments with heterogeneous concentrations of antibiotic stimulate the mutagenesis of bacterial populations, in addition to giving the necessary time to adapt to adverse conditions, which leads to super resistant mutant lineages that would not have been able to survive such antibiotic concentrations in homogeneous media.

Even when the current model was used to study a specific type of bacteria, the inputs can be changed to study a wide variety of bacteria if data is available.

As future work, the effects of others parameters can be explored, such as population density, mutation rate, the speed of expansion and spatial dimensions, and how all of them affect the way that bacteria mutates and forms a diversity of mutants with different capacities, making the model able to study other behaviors of interest in the scientific field regarding the evolution of bacteria, such as quorum sensing. Variables can be adjusted to model different bacterial populations in the same medium, obtaining a microbial competence model, where a colonization process could be observed in a microenvironment [9, 10]. Incorporation of the horizontal gene transfer phenomenon in bacterial populations to the model could enrich the analysis of AR acquisition. In addition, the simulation of the physical, chemical and biological processes that prevail in a bacterial community can be achieved, in order to predict the most suitable strain for the desired purpose [21]. This prediction could be applied in the decontamination of areas contaminated by pathogenic microorganisms, whether in hospital spaces or medical industrial environments, using appropriate strategies along adequate antibiotics, avoiding as much as possible the acquisition of AR.

Agent-based modeling results an adequate method for modeling biological complex systems such as the spatiotemporal bacterial evolution and AR development. It is important to continue the use of in silico simulations to understand and study the behavior and characteristics of complex microbial systems where in vitro experimentation are insufficient or expensive to generate new solutions. The use of in silico and in vitro experiments is a powerful combination to achieve a better understanding of specific phenomena.

References

1. Abar, S., Theodoropoulos, G.K., Lemarinier, P., O'Hare, G.M.: Agent based modelling and simulation tools: a review of the state-of-art software. Comput. Sci. Rev., 1–21 (2017)
2. Amigoni, F., Schiaffonati, V.: Multiagent-based simulation in biology: a critical analysis. In: Model-Based Reasoning in Science, Technology, and Medicine, pp. 179–191 (2007)
3. Banin, E., Hughes, D., Kuipers, O.P.: Editorial: bacterial pathogen, antibiotics and antibiotic resistance. FEMS Microbiol. Rev. **41**(3), 450–452 (2017)
4. Bauer, A.L., Beauchemin, C.A.A., Perelson, A.S.: Agent-based modeling of host-pathogen systems: the successes and challenges. Inf. Sci. **179**(10), 1379–1389 (2009)
5. Baym, M., Lieberman, T.D., Kelsic, E.D., Chait, R., Gross, R., Yelin, I., Kishony, R.: Spatiotemporal microbial evolution on antibiotic landscapes. Antibiot. Resist. **353**(6304), 1147–1152 (2016)
6. Bayrak, E.S., Wang, T., Jerums, M., Coufal, M., Goudar, C., Cinar, A., Undey, C.: In silico cell cycle predictor for mammalian cell culture bioreactor using agent-based modeling approach. IFAC-PapersOnLine **49**(7), 200–205 (2016)
7. Carley, K.M., Fridsma, D.B., Casma, E., Yahja, A., Altman, N., Chen, L.-C., Kaminsky, B., Nave, D.: Bio war: scalable agent-based model of bioattacks. Syst. Hum. **36**(2), 252–264 (2006)
8. Castro, C., Luz, D., David, F., Vásquez, C., Vargas, E., Gutiérrez, E., Franklin, L., Muñoz, M.: An agent-based model of the fission yeast cell cycle. Curr. Genet. (2018)
9. D'Acunto, B., Frunzo, L.: Qualitative analysis and simulations of a free boundary problem for multispecies biofilm models. Math. Comput. Model. **53**(9–10), 1596–1606 (2011)
10. D'Acunto, B., Frunzo, L., Klapper, I., Mattei, M.R.: Modeling multispecies biofilms including new bacterial species invasion. Math. Biosci. **259**, 20–26 (2015)
11. Duployez, C., Robert, J., Vachée, A.: Trimethoprim susceptibility in E. coli community-acquired urinary tract infections in France. Medecine et Maladies Infectieuses **48**(6), 410–413 (2018)
12. Gautam, S., Kalidindi, R., Humayun, M.Z.: SOS induction and mutagenesis by dnaQ missense alleles in wild type cells. Mutat. Res.—Fundam. Mol. Mech. Mutagen. **735**(1–2), 46–50 (2012)
13. Gilbert, N., Troitzsch, K.G.: Simulation for the Social Scientist (1999)
14. Gross, L.J.: Computer systems and models, use of. Encycl. Biodivers. **1**, 845–853 (2013)
15. Hegreness, M., Shoresh, N., Damian, D., Hartl, D., Kishony, R.: Accelerated evolution of resistance in multidrug environments. PNAS **105**(37), 13977–13981 (2008)
16. Hermsen, R., Deris, J.B., Hwa, T.: On the rapidity of antibiotic resistance evolution facilitated by a concentration gradient. PNAS **109**(27), 10775–10780 (2012)
17. Janion, C.: Some aspects of the SOS response system—a critical survey. **48**(3), 599–607 (2001)
18. Khataee, H.R., Aris, T.N.M., Sulaiman, M.N.: An agent-based model of muscle contraction process as a bio-robotic process. In: 2011 5th Malaysian Conference in Software Engineering. MySEC 2011, pp. 55–60 (2011)

19. Laxminarayan, R., Duse, A., Wattal, C., Zaidi, A.K., Wertheim, H.F., Sumpradit, N., Vlieghe, E., Hara, G.L., Gould, I.M., Goossens, H., Greko, C., So, A.D., Bigdeli, M., Tomson, G., Woodhouse, W., Ombaka, E., Peralta, A.Q., Qamar, F.N., Mir, F., Kariuki, S., Bhutta, Z.A., Coates, A., Bergstrom, R., Wright, G.D., Brown, E.D., Cars, O.: Antibiotic resistance-the need for global solutions. Lancet Infect. Dis. **13**(12), 1057–1098 (2013)
20. Lee, H., Popodi, E., Tang, H., Foster, P.L.: Rate and molecular spectrum of spontaneous mutations in the bacterium Escherichia coli as determined by whole-genome sequencing. Adsorpt. J. Int. Adsorpt. Soc. **109**(41), 2–4 (2012)
21. Mattei, M.R., Frunzo, L., D'Acunto, B., Esposito, G., Pirozzi, F.: Modelling microbial population dynamics in multispecies biofilms including Anammox bacteria. Ecol. Model. **304**, 44–58 (2015)
22. Murli, S., Walker, G.C.: SOS mutagenesis. Curr. Opin. Genet. Dev. **3**(5), 719–725 (1993)
23. Northrup, S.H., Erickson, H.P.: Kinetics of protein-protein association explained by Brownian dynamics computer simulation. Proc. Natl. Acad. Sci. **89**, 3338–3342 (1992)
24. Piddock, L.J.V.: Understanding drug resistance will improve the treatment of bacterial infections. Antimicrob. Resist. **15**, 639–640 (2017)
25. Scholar, E.: Trimethoprim. (1) (2007)
26. Tung, C.-k., Pourmand, N., Austin, R.H.: Connected microenvironments. Science, 1764–1767 (2011)
27. Wilensky, U.: NetLogo. Center for Connected Learning and Computer-Based Modeling, Northwestern University, Evanston, IL (1999)
28. Wilensky, U., Rand, W.: An Introduction to Agent-Based Modeling. MIT Press (2015)
29. Yu, J.S., Bagheri, N.: Multi-class and multi-scale models of complex biological phenomena. Curr. Opin. Biotechnol. **39**, 167–173 (2016)
30. Zhang, L., Wang, Z., Sagotsky, J.A., Deisboeck, T.S.: Multiscale agent-based cancer modeling. J. Math. Biol. **58**(4–5), 545–559 (2009)

Towards Intelligent Systems for Ubiquitous Computing: Tacit Knowledge-Inspired Ubicomp

Violeta Ocegueda-Miramontes, Mauricio A. Sanchez and Leocundo Aguilar

Abstract This work attempts to contribute to shortening the gap between the original conception of ubicomp and the current state of the art. Thus, the original concept has been analyzed to identify one of the main advances required to achieve this. Authors consider that context-awareness is the key element to accomplish this goal and compare the context-aware system architectures to humans' body parts, activities, and processes. Later they argue that, since context-aware systems are heavily influenced by humans context-awareness, it should implement techniques inspired by the human thinking process. Then, the concept of knowledge is reviewed altogether with a model of the human thinking process. Consequently, they propose the tacit context-modeling and the tacit context-reasoning techniques based on that model. Later, granular computing theory is proposed to implement the tacit context modeling technique, with which we propose the particular granular computing, the preconcept granular computing, and the concept granular computing. The authors conclude that the approach presented could help improve context-aware systems' performance since it takes inspiration on how they perceive the only perfect machine ever created works, the brain.

1 Introduction

The idea of an era of modern computing in which technology is bound up to everyday life in the smoothest way possible, able to adapt to different contexts, and capable of solving problems by themselves is nothing more than the philosophical inspiration

V. Ocegueda-Miramontes (✉) · M. A. Sanchez · L. Aguilar
Universidad Autónoma de Baja California, Calzada, Mexico
e-mail: violeta.ocegueda@uabc.edu.mx

Universidad #14418 Parque Industrial Internacional Tijuana, 22390 Tijuana, Mexico

M. A. Sanchez
e-mail: mauricio.sanchez@uabc.edu.mx

L. Aguilar
e-mail: laguilar@uabc.edu.mx

M. A. Sanchez et al. (eds.), *Applied Decision-Making*, Studies in Systems, Decision and Control 209, https://doi.org/10.1007/978-3-030-17985-4_4

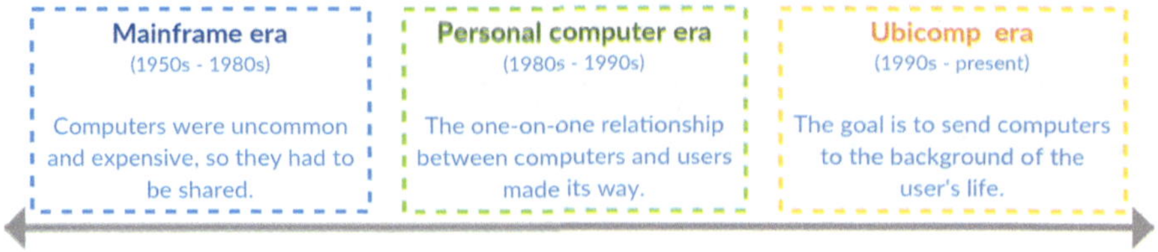

Fig. 1 Timeline of the modern computer's history

for ubiquitous computing (ubicomp for short) [40, 89, 94]. The concept-also referred to as "invisible computing" or "calm technology"—was coined by Mark Weiser at Xerox PARC in 1988 [40, 95], where he conceived a completely different approach in the way we interact with our surroundings. In his view, technological devices are in charge of making lives easier without interrupting our daily activities, making them unnoticeable by integrating them with already daily used objects, instead of being forced as a set apart activity.

To understand the motives behind Weiser's commitment to ubicomp, we should take a look at the history of the computer as we know it now, the modern computer, from which three main trends excel. In the first one (1950s–1980s), named the mainframe era, computers were so uncommon and expensive that they usually had to be shared, which gave way to a situation where many users made use of the same computer [96]. From the 1980s to 1990s, the price of computers lowered to reasonable prices, and that's how it was introduced to the market, resulting in many households acquiring it, what eventually headed towards a one-on-one relationship (one computer for one user); and that's how the Personal Computer (PC) era emerged [91]. Nevertheless, PCs were hard to use, demanding in time, and isolating. Hence, the third era surfaced (from the 1990s to present) in an attempt to amend these issues. This last era's philosophy is to take computers back to the background of the user's lives, allowing them to use computers without noticing them [40]. This trend is what we call the ubicomp era. Figure 1 sketches these three waves in the modern computer's history, along with their essential qualities.

However, the picture depicted above differs from the current trends in technology design. Nowadays, computers dominate our attention, detaching us from our surroundings and turning our existences solitary. What should be invisible technology has become invasive technology, to the point that some studies had been carried out to understand how technology overload has negatively impacted individuals' well being [4, 11, 14, 17, 28, 32, 35, 43, 46, 50, 51, 76–78, 84]. In this regard, Takayama [89] raises the need for a strategy to move in the direction of a new ubicomp that picks up its inceptive principals where computers deliver a first-person experience, making systems more calming, less distracting from social relationships, and less physically intrusive.

In an effort to understand what has to be done to achieve the so-called "invisibility" a further understanding of ubicomp is required. The next sections in this paper are organized as follows. In Sect. 2, we review the characteristics of ubicomp systems and its leading paradigms. We later delve into the context-awareness paradigm, in Sect. 3, as well as the main components of context-aware systems; and then present some related work. In Sect. 4, we present the proposal of this work, along with some needed concepts, such as knowledge, tacit knowledge, the tacit knowledge creation process, and granular computing, among others. Lastly, in Sect. 5, we withdraw the conclusions of the work, along with future work.

2 Ubiquitous Computing

For the sake of achieving ubicomp's ultimate purpose, we first need to comprehend the features that a computer or a system must possess for it to be recognized as a ubiquitous system.

2.1 Ubicomp Characteristics

There are different points of view concerning the characteristics an ubicomp system should bear, but we present some of those that we consider most relevant. Karvonen and Kujala [38] claim that ubicomp systems' primary features are context-awareness, natural interaction methods, invisibility, support for everyday tasks, and interconnectivity. Authors even go as far as to propose fulfillment levels for those characteristics—low, medium and high-, and explain how to apply this quality measurement to different technology devices. In [85, 86], researchers conducted a systematic review of the characteristics engineers should think about when developing ubicomp software projects; in this regard, they mention service omnipresence, invisibility, context sensitivity, adaptable behavior, experience capture, service discovery, function composition, spontaneous interoperability, heterogeneity, fault tolerance, scalability, quality of service, privacy, and trust.

On the other hand, Carvalho et al. [18–20] carried out an exhaustive literature review through which the author identified a list of 27 elements of ubicomp systems. The elements detected are acceptability, attention, availability, calmness, context-awareness, device capability, ease of use, effectiveness, efficiency, familiarity, interconnectivity, mobility, network capability, predictability, privacy, reliability, reversibility, robustness, safety, scalability, security, simplicity, transparency, trust, usability, user satisfaction, and utility. In this work, authors also observed that 5 out of the 27 characteristics listed had a higher frequency of appearance, rendering them as the most accepted features by literature as the ones a ubicomp system should possess (see Fig. 2). These five features are usability, context-awareness, transparency, privacy, and mobility. They also took the trouble to clarify what these terms stand for from their perspective:

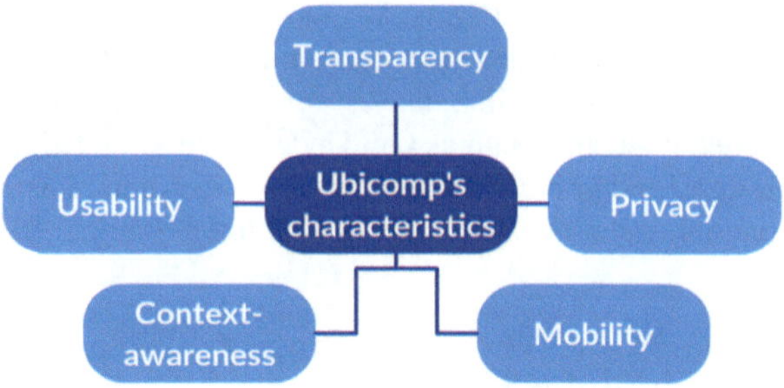

Fig. 2 Ubicomp's main characteristics according to literature

- **Usability**: denotes how easy it is for a user to understand, learn and use a system, and measures the extent to which that system is attractive to the user.
- **Context-awareness**: implies the capability to provide the user with relevant information, where relevancy depends on the user's task [90], and adapt the functionality of the system without distracting the user.
- **Transparency**: resorts to the ability to conceal a system from the user, so that he forgets he is using it.
- **Privacy**: has to do with the capability of a system to keep data safe.
- **Mobility**: represents the ability to maintain continuous access to data and resources, disregarding the location within a preset area.

To undertake the challenges the desirable characteristics of ubicomp systems presented to computing researchers, a set of paradigms developed along with the rise of ubiquitous computing.

2.2 Ubicomp Paradigms

Some of the most representative paradigms of ubiquitous computing we can name are wireless computing, intelligent environments, disaggregated computing, invisible computing, and wearable computing (see Fig. 3).

- **Wireless computing**: focuses on the problem of wireless communication, which is required by small computers that are expected to be carried [31, 81].
- **The intelligent environment**: targets spaces equipped with numerous networked devices, whether static or mobile, that interact with one another to provide users with access to information and services [8, 15, 44].
- **The disaggregated computing**: centers on physically separating a system into functional blocks and interconnecting them through high-speed networks [54, 63, 80].
- **Context-aware computing**: aims to build systems with the ability to gather useful information about the user and his current situation, and adapt its functionality, contents, and interfaces accordingly, without the user's explicit intervention [90].

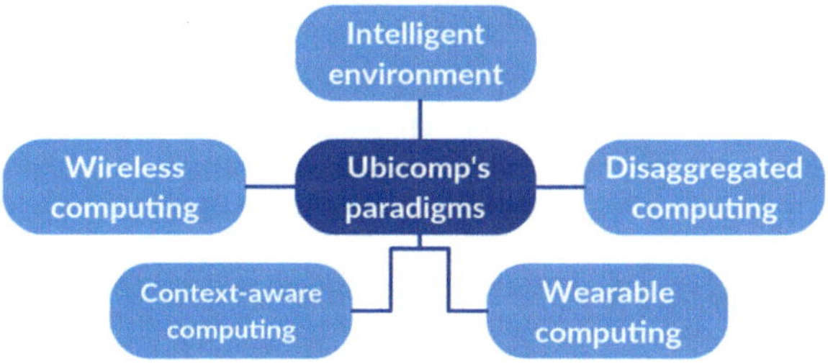

Fig. 3 Some of the most representative paradigms of ubicomp

- **Wearable computing**: is concerned with the seamless integration of portable devices into items of clothing without disrupting the wearer [88].

Among those paradigms, some researchers think of context-awareness as the essential element of ubicomp [52, 73, 90], because it's the one that addresses the gap between the user and the system understanding. Even Weiser himself stressed on the fact that computers should disappear from the user's awareness [93]:

> The most profound technologies are those that disappear. They weave themselves into the fabric of everyday life until they are indistinguishable from it.

He also stated that

> Only when things disappear in this way are we freed to use them without thinking and so to focus beyond them on new goals.

In his struggle to explain the "disappearance" phenomenon he resorted to already known concepts, such as "compiling" (Herbert A. Simon), the "tacit dimension" (Michael Polanyi), the "visual invariants" (J. J. Gibson), "horizon" and "ready-to-hand" (Hans Georg Gadamer and Martin Heidegger), "periphery" (John Seely Brown) [93].

The relevance of context-awareness surfaces due to the notion that humans convey ideas quite successfully. Trying to figure out how we make this possible, researchers identify three main factors influencing this result: the vast vocabulary of the language shared, the common conceptualization on the way the world operates, and an implicit realization of everyday situations [27]. In this way, when individuals talk to each other, they imply situational information that allows them to have a deeper understanding of what has explicitly been said with words (also known as the tacit understanding, according to Polanyi [70]). This is how researchers, such as Weiser, come up with the idea that systems could benefit from context-awareness.

In the next section, we elaborate on the concept of context awareness, its characteristics, and components.

3 Context-Awareness

Context-awareness has been widely explained within literature. Below, we present some of the definitions we found most interesting. Authors in [5, 16, 27] interpret it as the use of context to

> provide relevant information or services to the user, where relevancy depends on the user's task.

Khabou et al. [39] state that it is the capability of a system to

> adapt its behavior according to the location of use, the collection of nearby people, etc.

Not and Petrelli [61] declare that context awareness has to do with the capacity of a system to

> sense the current state of the environment and to respond accordingly.

Mylonas [55] defines it as the ability of a system to

> gather information about their environment at any given time and adapt their behaviors accordingly, i.e., provide services with full awareness of their current environment, which actually guides their responses.

Maran et al. [52] interpret it as the systems' capability to

> adapt and customize their execution based on information obtained from the environment.

For Riboni [73], it is the faculty of

> modeling and recognizing the user's context [...] to effectively adapt applications and services to his/her needs, goals and expectations.

In their work, Temdee and Prasad [90] describe it as the capacity of a system to

> adapt its operation or behavior to the current contexts with or without the explicit intention of the user intervention.

Machado et al. [49] explain context-awareness as the ability of a system to

> adapt their behaviors to the changing environment with minimal human intervention.

Börger and Raschke [12] define it as the capability to

> specify and manage 'program environments' which influence the behavioral meaning of the program, i.e., determine the result of its execution in dependence of a given environment.

Liu et al. [48] explain it as what

> enables users to enjoy services anytime and anywhere, but also ensures that mobile platforms can continuously receive users' contexts, be aware of and predict users needs, and meanwhile search relevant information for users.

Grubert et al. [33] depict it as the

> facility to establish context.

Yürür et al. [97] define it as the ability to

> discover and react to changes in the environment [devices] are situated in.

For Hofer et al. [36], it is the capability to

> provide services not only to people at any time, anywhere, with any media but specifically to communicate the right thing, at the right time in the right way.

Couderc and Kermarrec [24] visualize it as the capability to be

> closely integrated into the real world to implicitly assist people actions.

Schmidt et al. [82] define context-awareness as

> knowledge about the user's and IT device's state, including surrounding, situation, and to a lesser extent, location.

Pascoe et al. [65] picture it as the ability of a system to

> sense, react or adapt to its environment of use.

Finally, Abowd et al. [2] describe context-awareness as the use of context to

> provide task-relevant information and/or services to a user.

In Table 1, we present a comparison between the context-awareness different definitions we found in the literature to facilitate the identification of its attributes.

On the other hand, Table 2 shows an overview of the most mentioned characteristics we found in literature about context-awareness.

Considering the definitions found in literature and the features conferred by different authors, we interpret context-awareness as the capability of a system to:

1. Sense and gather information from the current and changing environment in order to
    - specify, manage, adapt to, and respond accordingly to such an environment.
2. Model and recognize the user's context, and predict the user's needs, without relying on the user's explicit intervention, aiming to
    - adapt to the user's needs, goals and expectations,
    - provide the user with relevant and timely information and services, where relevancy depends on the user, with a pertinent representation, anywhere/anytime,
    - implicitly assist the user's actions.
3. Smoothly integrate into the environment.

In Fig. 4 we schematize our interpretation of context-awareness, considering the attributes pointed out by the different authors.

As we may see, to build a system that meets the criteria for context-awareness is not an easy task, but a quite challenging one. Still, identifying the components of such a system may be the right step to follow.

Table 1 Context-awareness' attributes found in literature

References	Definition	Characteristics
[5, 16, 27]	"provide relevant information or services to the user, where relevancy depends on the user's task"	• Provides relevant information/services to the user •Relevancy depends on the user's task
[39]	"adapt its behavior according to the location of use, the collection of nearby people, etc".	• Adapts according to location, nearby people, etc.
[61]	"sense the current state of the environment and to respond accordingly"	• Senses environment • Responds according to environment
[55]	"gather information about their environment at any given time and adapt their behaviors accordingly, i.e., provide services with full awareness of their current environment, which actually guides their responses"	• Gathers information • Adapts to environment • Provides services • Responds according to environment
[52]	"adapt and customize their execution based on information obtained from the environment"	• Adapts to environment • Customizes execution to environment
[73]	"modelling and recognizing the user's context ... to effectively adapt applications and services to his/her needs, goals and expectations"	• Models and recognizes the user's context • Adapts to the user needs, goals and expectation
[90]	"adapt its operation or behavior to the current context with or whithout the explicit intention of the user intervention"	• Adapts to current context Does not rely on the user's explicit intervention
[49]	"adapt their behaviors to the changing environment with minimal human intervention"	• Adapts to changing environment • Minimal human intervention
[12]	"specify and manage'program environments' which influence the behavioral meaning of the program, i.e. determine the result of its execution in dependence of a given environment".	• Specifies and manages environment • Changes behavior according to environment
[48]	"enables users to enjoy services anytime and anywhere, but also ensures that mobile platforms can continuously receive users' contexts, be aware of and predict users' needs, and meanwhile search relevant information for users".	• Enables services anytime/anywhere • Receives user's context • Predicts user's needs • Searches for relevant information for users
[33]	"facility to stablish context"	• Establishes context
[97]	"discover and react to changes in the environment [devices] are situated in".	• Discovers environment • Reacts to environment
[36]	"provide services not only to people at any time, any where, with any media but specifically to communicate the right thing, at the right time in the right way"	• Provides services at anytime/anywhere • Communicates the right thing at the right time in the right way
[24]	"closely integrated into the real world to implicitly assist people actions"	• Integrates into the real world • Assists people actions implicitly
[82]	"knowledge about the user's and IT device's state, including surrounding, situation, and to a lesser extent, location"	• Knowledge about the user and device
[65]	"sense, react or adapt to its environment of use"	• Senses environment • Reacts to environment • Adatps to environment
[2]	"provide task-relevant information and/or services to a user"	• Provides task-relevant information/services

Table 2 Overview of the most mentioned attributes found in literature about context-awareness

Characteristics	References
Provides relevant information/services to the user	[2, 5, 16, 27, 48, 55]
Relevancy depends on the user's task	[5, 16, 27]
Adapts to current context	[39, 52, 55, 65, 90]
Senses and gathers information about the environment	[48, 55, 55, 61, 65, 97]
Responds according to environment	[12, 52, 55, 61, 65, 97]
Models and recognizes the user's context	[73, 82]
Adpats to the user's needs, goals and expectations	[73]
Does not rely on the user's explicit intervention	[49, 90]
Adapts to changing environment	[49]
Specifies and manages environment	[12, 33]
Enables services anytime/anywhere	[36, 48]
Communicates the right thing at the right time in the right way	[36]
Predicts user's needs	[48]
Integrates into the real world	[24]
Assists people actions implicitly	[24]

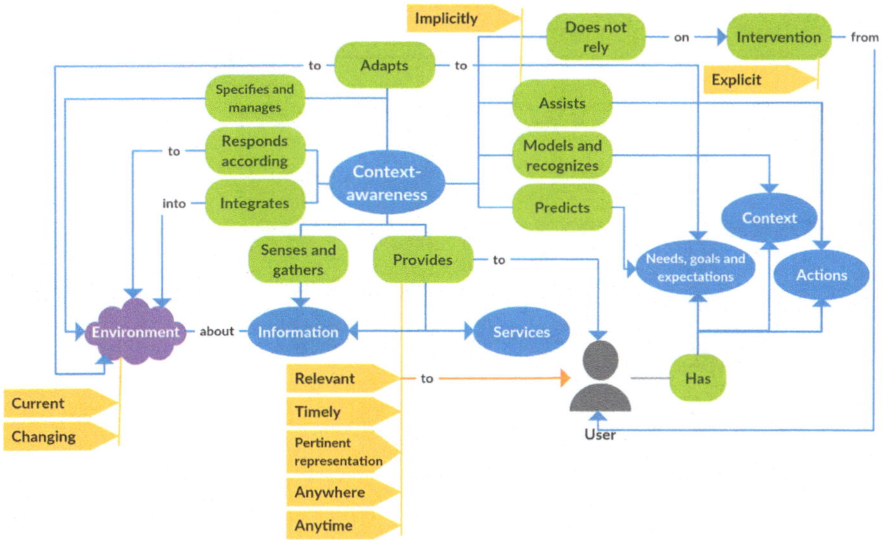

Fig. 4 Interpretation of context-awareness

3.1 The Components of Context-Aware Systems

Some researchers have suggested different architectures for context-aware systems [5, 7, 22, 29, 39, 90]. In Fig. 5, we show some of the latest suggested architectures and highlight how their components are related to one another. As can be noticed, these architectures display three common components:

- **The perceiving component**: is in charge of acquiring and storing data from the physical world to be used by the system to make decisions.
- **The thinking component**: this component contains two phases, context modeling and context reasoning. Context modeling refers to the process of defining a formal, static or dynamic, representation of the context. Context reasoning is the process of establishing a methodology to infer new knowledge to gain a better understanding of the context.
- **The acting component**: once data is analyzed and made sense of, this component takes the appropriate actions, either to communicate information or to provide services.

According to [40] the most relevant issue to consider when designing a context-aware systems is to clearly define the knowledge the system will need of the physical world and the user's behavior, how the user might influence the system; besides determining the mechanisms to represent and maintain such knowledge, and how to recover from wrong conclusions. That is, the most significative element of the context-aware system's architecture is the thinking component. Therefore, in this work, we focus on the thinking phase, which involves context modeling and context

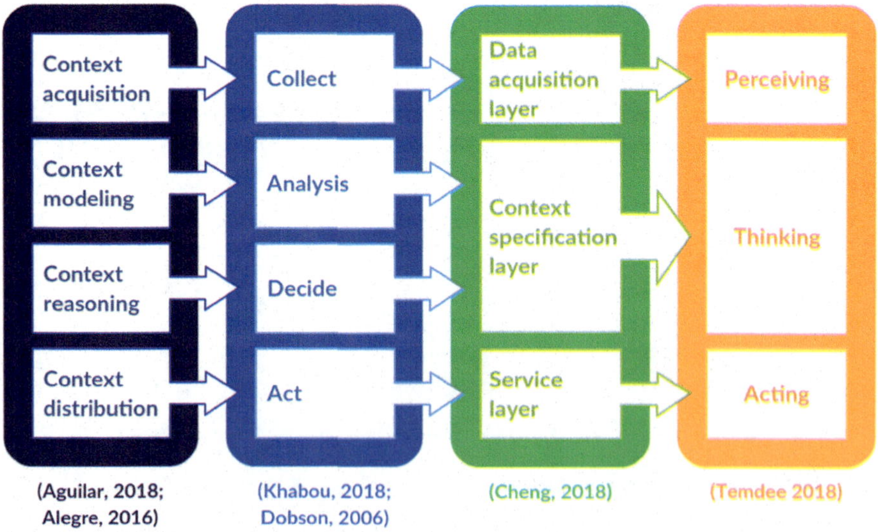

Fig. 5 The common components of context-aware systems' architectures

reasoning. Next, let's take a look at the techniques used to implement these two processes.

Context modeling techniques
Some of the most employed techniques in context modeling are [74, 90]: key-value modeling, markup scheme modeling, graphical modeling, object-based modeling, logic-based modeling, and ontology-based modeling. Following, we briefly describe these techniques. Key-value modeling uses key-value pairs to model context information in different formats. Markup scheme modeling uses tags to model data. Graphical modeling captures the relationships within a context. Object-based modeling uses encapsulation and inheritance to represent context in code format. Logic-based modeling uses facts, expressions, and rules to define the context model. Ontology-based modeling uses semantic technologies to describe the taxonomies of concepts and relationships.

Context reasoning techniques
On the other hand, context reasoning techniques can be classified into six categories [74, 90]: rules, fuzzy logic, ontology-based reasoning, probabilistic logic, supervised learning and unsupervised learning. Next, we outline the techniques mentioned.

A rule is a structure in the IF-THEN-ELSE format used to generate high-level context information from low-level context. Fuzzy logic is a decision-making technique based on imprecise information, by using logical expressions that describe memberships in fuzzy sets. Ontology-based reasoning is a technique based on description logic. Probabilistic logic depends on logic and probability to handle uncertainty. Supervised learning gathers and labels training data according to expected targets or results; then, a learning algorithm infers a function that maps an input to an output based on the training data. This function is called the generalization function and is designed to deliver reasonable outputs for any inputs that are not in the training data. Some supervised learning techniques are decision trees, bayesian networks, artificial neural networks, and support vector machines. Unsupervised learning uses training data that has not been labeled and searches for similarities in the data and reacts based on the absence or presence of such similarities in new data. Some techniques that implement unsupervised learning are the k-nearest neighbor and the kohonen self-organizing map.

The upcoming step is to review the different implementations of the thinking component for context-aware systems.

3.2 Related Work

Lee and Helal [45] use stochastic analysis of sensor data to model context. Such context models are generated with supervised learning, which operates based on three different methods: (1) Conditional Probability Table (CPT) is used to define the number of contexts; (2) K-means clustering facilitates the formation of groups, each of which is later claimed as a context; (3) Principal Component Analysis (PCA) finds the most representative sensors in each context. These three methods give as a result the complete context model. The work concludes that the methodology proposed has a good performance, besides it helps to reduce the human effort to process context recognition while increasing its automaticity.

In [47], authors present an adaptive security-related data collector for heterogeneous network context. The collector addresses the problem of heterogeneity of the network system by integrating a Security-related Data Description Language (SDDL), which describes the security-related data to facilitate data collection and processing. It also employs adaptive sampling algorithms to enhance collection efficiency and accuracy, and to reduce the amount of collected data. For the latter, they combine three statistical sampling algorithms—simple random sampling, systematic sampling, and poisson sampling—to propose the Adaptive Security-related Data Determination based on Network Context (ASDDNC) algorithm.

Zheng et al. [98] suggest the integration of context-aware recommender systems (CARS) with multi-criteria recommender systems (MCRS) into a recommender system and experiment with educational data and tourism data. On the side of the CARS, they select the context-aware matrix factorization (CAMF) for context modeling to produce context-aware recommendations. Then, they incorporate context into their multi-criteria rating predictions and rating aggregations. Finally, they state that the process outlined may boost the accuracy of recommendations.

In [5], researchers present the CAMeOnto ontology that captures generic concepts to a higher level by providing mechanisms to extend the specific information of a context in a hierarchical manner. The ontology's design is based on the 5Ws principles: who, when, what, where and why. It categorizes context into internal context, external context, and boundary context. Besides, it has two hierarchical levels. The first one is a general and independent domain ontology, and it has six contextual classes: user, activity, time, device, services, and location. These classes have a correspondence to the 5W's: the user class answers to the question of "who", time defines "when", location specifies "where", services class identifies"what" and, device and activity describe the "why". The second level in the ontology enables the addition of specific domain context ontologies. Authors also present case studies for a smart classroom, a surgery room, and a semantic workplace learning framework, for which the CAMeOnto ontology successfully modeled the context, according to them.

Ruta et al. [75] approach context-awareness in Vehicular Ad-hoc Networks (VANETs) by introducing a logic-based framework that enables data management and mining in VANETs. The distributed collaborative framework is built from (1) a non-monotonic inference algorithm for knowledge fusion, with inconsistency and resilience troubleshooting to manage false information; and, (2) a dissemination protocol to bidirectionally propagate knowledge in VANETs.

Cheng et al. [22] conceive a service-oriented framework for context-awareness that contemplates data acquisition, context specification, and service. In this framework, the specification layer takes care of the context modeling and the context-reasoning processes. For context modeling, they use ontology models. In context reasoning, they describe two steps: first they use ontology-based reasoning for low-level context; then, they implement high-level inference rules for high-level context.

In [39], authors introduce an analysis module to support application designers in the design and development of context-aware applications. The analysis module is constituted by two main tasks: an analysis task for context change detection and an analysis task for context prediction. The former task has five steps: (1) the analysis parameter specification, that has to do with specifying the parameter to be analyzed; (2) the parameter category specification, where they define categories according to the context parameter evolution; (3) thresholds definition according to the user needs; (4) mathematical models definition, in this step they define different mathematical models such as the constant model, the step function model, the moving average model, the simple moving average model, the weighted moving average model, the exponential weighted moving average and the application designer specific model; and, (5) the notification release technique definition, in which the application designer chooses the way to hand over notifications.

Finally, Machado et al. [49] propose the Hybrid Architecture to Correlate Contextual Data (HACCD), which integrates hybrid context modeling. This architecture has six layers. The first one is the medium access layer for context acquisition. The second layer is the preprocessing one that normalizes and creates new data. The third one is the processing layer that implements hybrid reasoning for context processing. The repository communication manager is the fourth layer, and it provides methods for search, insertion, and deletion of data. The storage layer is the fifth one, and it offers three storage models—a relational model, a non-relational model, and a triple model. Lastly, the correlation layer is the sixth one, and it uses the compositional rules strategy to correlate the data stored in distinct models. In the processing layer of their HACCD architecture, they use three strategies for context reasoning: rules, ontology and supervised learning. Rules are used to correlate events to search for patterns; ontology is used in two forms, (1) internal axioms are used to infer knowledge about the classes that assemble the ontology, (2) external rules are applied to enable the system to infer new context information about the ontology instances; (3) for supervised learning is implemented with decision trees. As results, they reported that the correlation of data from different natures might be used to improve decision-making.

As can be seen from the works described above, context modeling and context reasoning techniques currently found in literature have greatly contributed to reaching the context-awareness ultimate goal. However, despite all the effort invested, there's still a long road to success. We consider that an essential issue is that these techniques have left aside the fact that context-awareness was inspired by humans' capacity to grasp and abstract the real world. For that matter, context-awareness is not only inspired by human beings, but it actually is an intrinsically human characteristic. Such is the case that there can be observed a close connection between the common components of context-aware systems' architectures to human's body parts or processes. This connection is more clearly depicted in the architecture presented by [90] (see Fig. 6), which breaks down into three parts: perceiving, thinking and acting. The perceiving component in the context-aware system refers to the sensors or any data input mechanism the system may have, many of which could have an almost direct relation to the human senses. The thinking component is where context modeling and context reasoning happen in the context-aware system, and humans

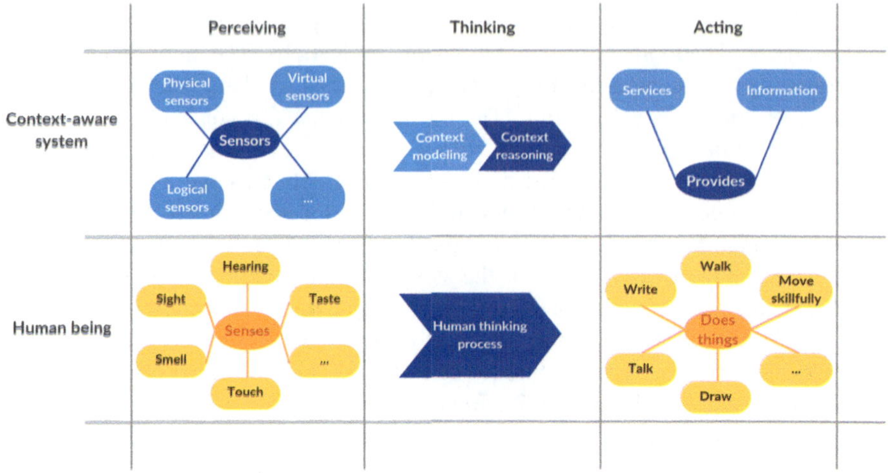

Fig. 6 Context-aware system vs human context-awareness

indeed possess such processes which occur in the brain. The only difference is that the processes in the brain are undeniably far more sophisticated since the brain is a considerably more powerful machine. Finally, after acquiring information from the environment, modeling the data to build the context model, and reasoning according to that model, the acting component in the context-aware system reacts to the context by providing services and information to the user. Analogously, humans react to the context by walking, writing, talking, drawing, and skillfully executing tasks, among other things.

In the next section, we propose a context-modeling technique that takes into consideration the way humans process knowledge about the environment.

4 Tacit Knowledge in Context Awareness

To summarize the previous section, context-aware systems pursue to simulate human behaviors. Thereupon, when designing those behaviors, we should try to stick as much as possible to the way humans perform them. For example, for the thinking component, we could take into consideration how humans construct knowledge about their context and how they make decisions based on it. To achieve this, it is important to understand what knowledge is.

4.1 Knowledge

Knowledge is a controversial concept that has been broadly redefined through time by many researchers such as Polanyi [70], Nonaka [57], Davenport [26], Baumard [10], Johannessen [37], and Collins [23], inter alia. But Nonaka [60] presents an interesting definition by stating that knowledge is

> information in context and once we add context we add tacitness.

Though they have not reached an agreement and almost every researcher has his own interpretation on what the concept involves, there is a consensus that knowledge is a two-sided coin, which has an explicit and a tacit side [3, 6, 13, 25, 30, 34, 41, 42, 53, 56, 58, 59, 64, 69, 72, 79, 83, 87, 92]. This duality is conceived by Polanyi in his book "The Study of Man" [69], where he defines explicit knowledge (EK) as the knowledge

> set out in written words or maps, or mathematical formulae

and tacit knowledge (TK) as

> unformulated knowledge, such as we have of something we are in the act of doing.

For Polanyi, these two types of knowledge are not mutually exclusive, but EK relies on TK to be tacitly understood and applied, as he argues that [69]

> all knowledge rests on understanding.

The essential difference between TK and EK is the kind of awareness to which each one is related. Awareness is strongly linked to self-reflection or inner search [1], which is divided into two types: subsidiary and focal awareness, connected to TK and EK, respectively [69]. The focal awareness enables critical reflection on knowledge. Conversely, knowledge in the subsidiary awareness cannot be focused on, ergo it cannot be critically reflected on, and the only way to access it is through pointers to a meaning.

Consequently, though TK is inherently present in all knowledge, it cannot be focused on. But for EK to be considered knowledge it has to be linked to the subsidiary awareness, where TK dwells [71]. Thus all knowledge is either TK or grounded on it, and completely EK is not possible [69]:

> nothing that is said, written or printed, can ever mean anything in itself: for it is only a person who utters something - or who listens to it or reads it- who can mean something by it.

In a nutshell, all knowledge is rooted in TK; that is, understanding depends on the TK an individual possesses. Therefore, TK is what enables humans to make sense of things by grasping context. Hence, we propose that TK should be considered in the scheming of the thinking component of context-aware systems, both in context modeling and context reasoning.

Once established that TK should play a role within context-aware systems, the next step is to figure out how this type of knowledge works. Ergo, we should follow by describing how it is built.

4.2 Tacit Knowledge Elements

In [62] we present the idea that knowledge in the brain has three main elements: "particulars", "preconcepts", and "concepts".

The term particular is introduced by Polanyi in [70]. Even though he does not provide a precise definition, he yields the concept through multiple explanations, from which we conclude that particulars do not hold a meaning by themselves. Instead, the concept they shape is what encloses a significance. As stated by Polanyi [69], particulars cannot be reflected on due to they are situated behind the barrier of language. In this sense, even when humans use particulars for reasoning, they can never be aware of them. However, knowledge cannot be built without particulars that support its understanding, which leads us to infer that the human brain relies on its already acquired particulars to provide information with a meaning. Along these lines, we understand particulars as the basis of all knowledge, that does not convey a meaning by themselves and can only hold significance when linked to more particulars.

On the other hand, the simplest preconcepts result from linking particulars in the brain. Preconcepts may manifest an increasing complexity as those connect to ever more particulars, and eventually to other preconcepts. Like particulars, preconcepts cannot be made sense of, but they are closer to what we know as concepts.

The exact moment in which a set of linked preconcepts become a concept is still unknown, but we believe that it just somehow happens. Therefore, we conceive a concept as an emergent property that results from the association of a vast amount of preconcepts, which in turn are formed by particulars. In Fig. 7, we depict how we believe particulars, preconcepts, and concepts intertwine. Besides, Fig. 8 shows how these three conceptions are situated in awareness.

Fig. 7 Particulars, pre-concepts and concepts

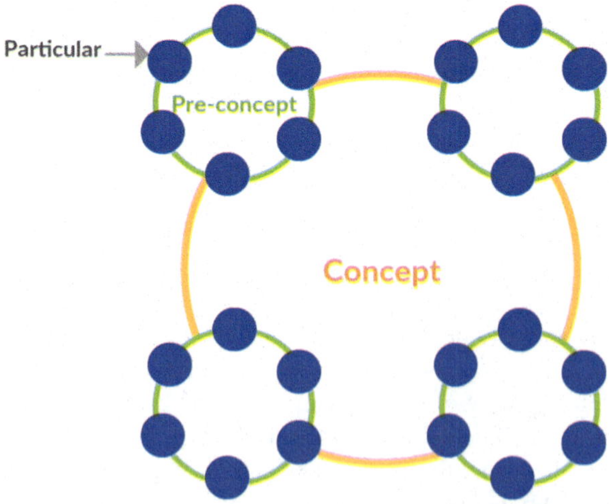

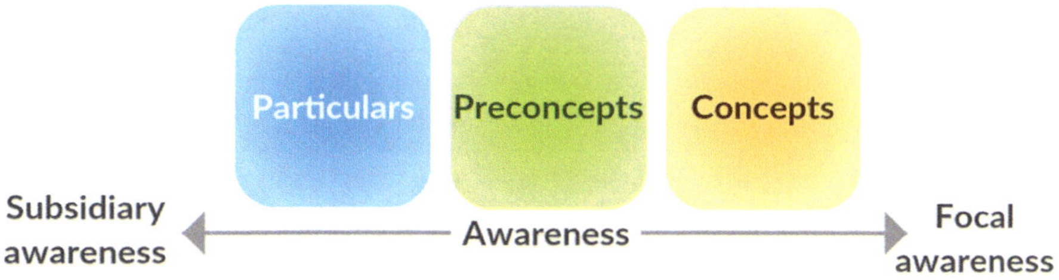

Fig. 8 Particulars, pre-concepts and concepts situated in awareness

In order to understand how these concepts interact to construct knowledge, in the next section, we pick up the tacit knowledge's acquisition process presented in [62] but focus solely on the knowledge creation process.

4.3 Tacit Knowledge Creation Process

The thinking process in humans begins when the sensory organs send stimuli to the brain in the form of impulses. The brain decodes those impulses and creates synapsis. Synapses in the brain are created/inhibited by experiencing the world [21]. When the brain receives the same information over and over, the synapses that information is related to is reinforced, creating stronger synapses. If, by contrast, the brain stops receiving certain information, the synapses related to it become weaker and disappear with time.

The synapses creation process generates synaptic patterns, which are the interpretations the brain gives to the sensory input, that the sensory organs provide. Synaptic patterns engender what we know as particulars, meaning that a specific arrangement of synapses results into peculiar particulars. Connections between these synaptic patterns or particulars, somehow give rise to preconcepts. A growing associating of preconcepts to other preconcepts and particulars eventually provokes a qualitative change that causes the emergence of a concept. If the synaptic patterns that hold a concept change, then the concept changes. The formation of synapses between the synaptic patterns of concepts generates increasingly complex concepts. Figure 9 portrays an interpretation of the human thinking process.

For the procedure described, we consider particulars, preconcepts, and concepts to be emergent properties of synapses connections, where:

- A particular is knowledge's minimum unit that dwells in the subsidiary awareness and cannot be taken to focal awareness.
- A preconcept is a piece of knowledge that is neither a particular nor a concept, but something in between. As for particulars, it resides in the subsidiary awareness and cannot be brought into focal awareness.
- A concept is a piece of knowledge that can be made sense of and can be held by focal awareness.

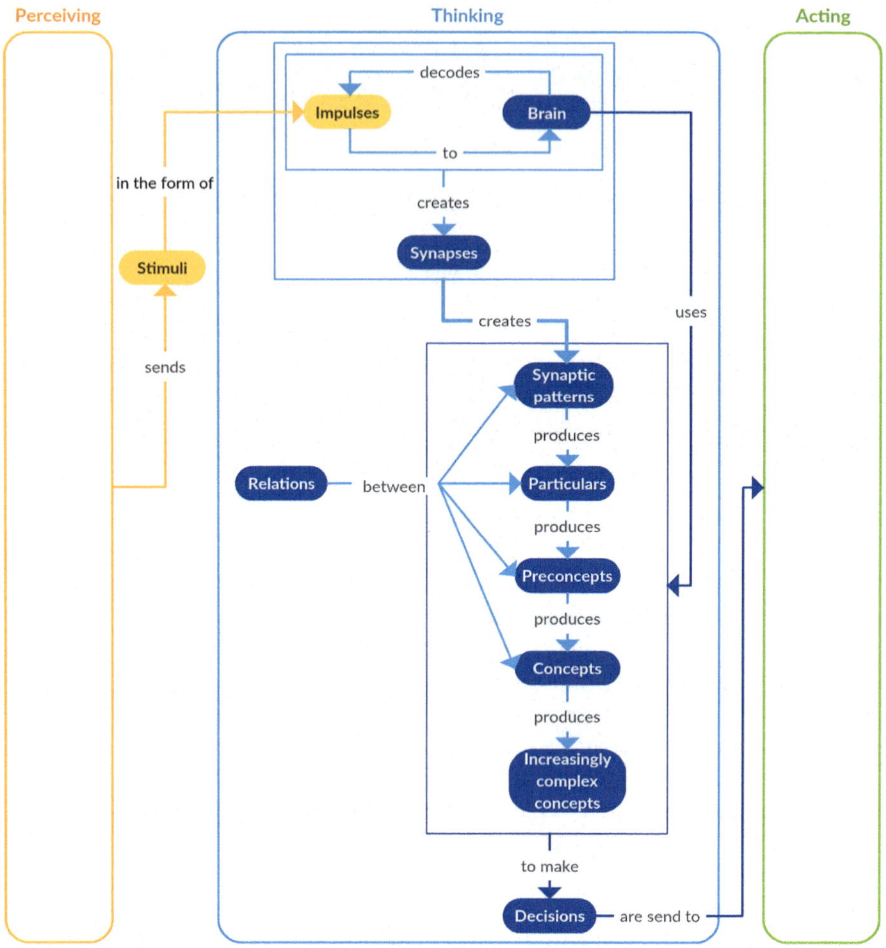

Fig. 9 Sketch of the human thinking process

4.4 *Proposal*

Based on all that is stated in this work, we propose to take inspiration in the TK's creation process to design techniques for context modeling and context reasoning to try to replicate the human mental processes. The tacit context modeling technique pursues to imitate humans' knowledge creation process, while the tacit context reasoning technique would approach the decision-making problem. Figure 10, we situate these techniques within the human being's thinking process. However, in this work, we focus solely on the design of the tacit context modeling technique.

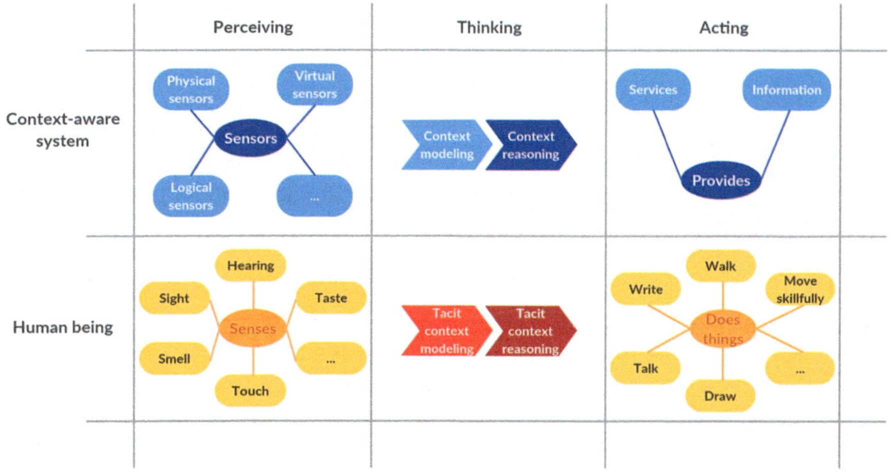

Fig. 10 Tacit context modeling and tacit context reasoning for human context-awareness

4.5 Tacit Context Modeling

The tacit context modeling technique we propose resorts to granular computing theory to produce particulars, preconcepts, and concepts from data. Figure 11 outlines the way in which these concepts interweave. At this point, it would be helpful to explain what this theory involves.

Granular computing consists of a collection of conceptions, techniques, algorithms, and applications that aim to process information granules [67]. An information granule is defined as [9]

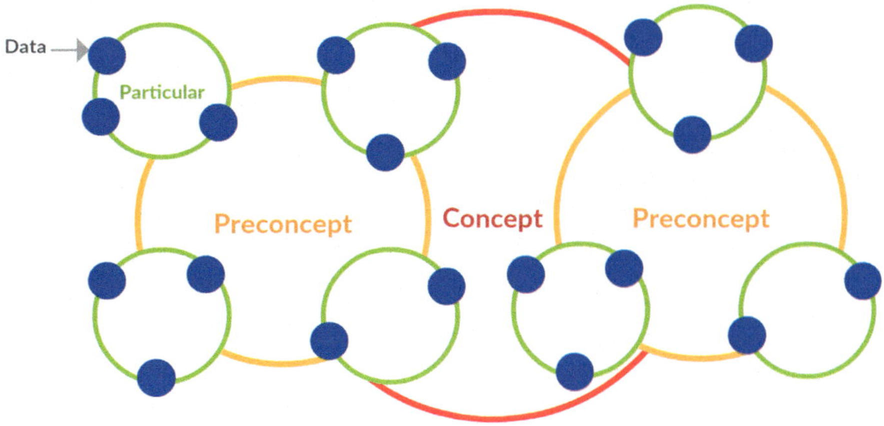

Fig. 11 Tacit context modeling

a semantically meaningful grouping of elements on their indistinguishability, similarity, proximity or functionality.

This theory uses granules such as classes, clusters, subsets, groups, and intervals to construct computational models. To form granules, it relies on the principle of justifiable granularity that targets the production of a meaningful representation based on numeric data, $D = \{x_1, x_2, ..., x_N\}$ [68]. The granulation of information is the process through which a transformation function g operates on a set of data $D = \{x_1, x_2, ..., x_N\}$, giving as a result an information granule G:

$$g : \{x_1, x_2, ..., x_N\} \rightarrow G \tag{1}$$

The central axis in this process is the design of the transformation function g and the construction of an optimization criterion, which determines the resulting information granule G. Regarding the optimization criteria, there are two antagonistic requirements. The first one has to do with generating a complex enough information granule, in such a way that it represents as much as possible the data gathered. The second one, is about preserving high specificity of the information granule. To build the information granule Ω, that complies with the requirements described, we have to follow two steps [66]:

- **Step one**: Determine the numeric representative of the data D. The median, mean or the standard deviation can be used as good representatives for samples of numeric data.
- **Step two**: Define the lower and upper bounds of the information granule Ω, denoted by a and b, respectively. This numeric interval must satisfy two conditions

1. it should accumulate the largest quantity of data possible within the bounds of Ω;
2. the support of Ω should be as low as possible, to allow Ω to be detailed enough

To address the conflict between the experimental evidence and the specificity requirements, [68] uses the maximization of the product

$$V = f_1 * f_2 \tag{2}$$

where:

f_1 is an increasing function that counts the number of data falling within the bounds of Ω. To do it, it takes into consideration the cardinality of Ω, $card\{x_k \in \Omega\}$. f_1 can be stated as $f_1(card\{x_k \in \Omega\})$. In the simplest case scenario, this delivers an identity function, $f_1(u) = u$.

f_2 is a decreasing function that measures the specificity of the information granule, by using the length of the interval Ω. For a higher specificity, a lower value of $f_2(length(\Omega))$ is desirable. f_2 takes the form $f_2(u) = \exp(-\alpha u)$, where α is a positive value that helps to calibrate the impact of the specificity criterion on the information granule, which offers flexibility in the delivered information granule. Bigger values of α boosts the relevance of the specificity criterion.

Pedrycz [68] exemplifies the construction of an information granule Ω as follows:

1. He stablises the median, denoted by m, as the numeric representative of the data.
2. The upper bound of Ω is determined.
    - To quantify the specificity of the information granule, the length of Ω is established as $|m(D) - b|$.
    - The decreasing function f_2 is described as $f_2(|m(D) - b|)$.
    - The cardinality function f_1 counts the number of elements of the data situated to the right from the median, $card\{x_k \in \Omega, x_k > m(D)\}$.
    - $f_1(card\{x_k \in \Omega, x_k > m(D)\}$ is computed.
    - The maximization of the product $V(b_{opt}) = max_{b>m(D)} V(b)$ is calculated.
3. The lower bound of Ω is decided.
    - The length of Ω is enacted as $|m(D) - a|$.
    - The decreasing function f_2 is put in the form $f_2(|m(D) - a|)$.
    - The cardinality function f_1 counts the number of elements positioned to the left from the median, $card\{x_k \in \Omega, x_k < m(D)\}$.
    - $f_1(card\{x_k \in \Omega, x_k < m(D)\}$ is estimated.
    - The maximization of the product $V(a_{opt}) = max_{a<m(D)} V(a)$ is carried out.

Now, for the tacit context modeling, we suggest to make particulars, preconcepts and concepts information granules. For that matter we define the maximization products for each type of granule:

- For particulars:

$$P = p_1 * p_2 \tag{3}$$

- For preconcepts:

$$PC = pc_1 * pc_2 \tag{4}$$

- For concepts:

$$C = c_1 * c_2 \tag{5}$$

Next section exhibits how information granules would be computed for each kind of granule.

4.5.1 Particular Granule Computing

The process to calculate the particular granule is described below:

1. The median, denoted by m_p, is used as the numeric representative of the data.
2. The upper bound of the interval is determined.

- The length of Ω_p is $|m_p(D) - b_p|$.
- The decreasing function p_2 is $p_2(|m_p(D) - b_p|)$.
- The cardinality function p_1 counts the number of elements of the data situated to the right from the median, $card\{x_k \in \Omega_p, x_k > m_p(D)\}$.
- $p_1(card\{x_k \in \Omega_p, x_k > m_p(D)\}$ is computed.
- The maximization of the product $P(b_{p-opt}) = max_{b_p > m_p(D)} P(b_p)$ is calculated.

3. The lower bound of Ω_p is decided.

- The length of Ω_p is $|m_p(D) - a_p|$.
- The decreasing function p_2 is $p_2(|m_p(D) - a_p|)$.
- The cardinality function p_1 counts the number of elements of the data positioned to the left from the median, $card\{x_k \in \Omega_p, x_k < m_p(D)\}$.
- $p_1(card\{x_k \in \Omega_p, x_k < m_p(D)\}$ is estimated.
- The maximization of the product $P(a_{p-opt}) = max_{a_p < m_p(D)} P(a_p)$ is carried out.

4.5.2 Preconcept Granule Computing

To calculate preconcept granules we use the particular granules generated from the previous step:

1. The median, denoted by m_{pc}, is used as the numeric representative of the particulars.
2. The upper bound of the interval is determined.

- The length of Ω_{pc} is $|m_{pc}(P) - b_{pc}|$.
- The decreasing function pc_2 is $pc_2(|m_{pc}(P) - b_{pc}|)$.
- The cardinality function pc_1 counts the number of particulars situated to the right from the median, $card\{y_k \in \Omega_{pc}, y_k > m_{pc}(P)\}$.
- $pc_1(card\{y_k \in \Omega_{pc}, y_k > m_{pc}(P)\}$ is computed.
- The maximization of the product $PC(b_{pc-opt}) = max_{b_{pc} > m_{pc}(P)} PC(b_{pc})$ is calculated.

3. The lower bound of Ω_{pc} is decided.

- The length of Ω_{pc} is $|m_{pc}(P) - a_{pc}|$.
- The decreasing function pc_2 is $pc_2(|m_{pc}(P) - a_{pc}|)$.
- The cardinality function pc_1 counts the number of particulars positioned to the left from the median, $card\{y_k \in \Omega_{pc}, y_k < m_{pc}(P)\}$.
- $pc_1(card\{y_k \in \Omega_{pc}, y_k < m_{pc}(P)\}$ is estimated.
- The maximization of the product $PC(a_{pc-opt}) = max_{a_{pc} < m_{pc}(P)} PC(a_{pc})$ is carried out.

4.5.3 Concept Granule Computing

Finally, in the process to calculate concept granules we use the preconcept granules as inputs:

1. The median, denoted by m_c, is used as the numeric representative of the preconcepts.
2. The upper bound of the interval is determined.
   - The length of Ω_c is $|m_c(PC) - b_c|$.
   - The decreasing function c_2 is $c_2(|m_c(PC) - b_c|)$.
   - The cardinality function c_1 counts the number of preconcepts situated to the right from the median, $card\{z_k \in \Omega_c, z_k > m_c(PC)\}$.
   - $c_1(card\{z_k \in \Omega_c, z_k > m_c(PC)\}$ is computed.
   - The maximization of the product $C(b_{c-opt}) = max_{b_c > m_c(PC)} C(b_c)$ is calculated.
3. The lower bound of Ω_c is decided.
   - The length of Ω_c is $|m_c(PC) - a_c|$.
   - The decreasing function c_2 is $c_2(|m_c(PC) - a_c|)$.
   - The cardinality function c_1 counts the number of preconcepts positioned to the left from the median, $card\{z_k \in \Omega_c, z_k < m_c(PC)\}$.
   - $c_1(card\{z_k \in \Omega_c, z_k < m_c(PC)\}$ is estimated.
   - The maximization of the product $C(a_{c-opt}) = max_{a_c < m_c(PC)} C(a_c)$ is carried out.

Figure 12 depicts the process through which data transforms into concepts. In this procedure, the data collected by the perceiving component of the context-aware system is the input of the thinking component. The latter component takes the data and sends it to the tacit context modeling module, which hands over this data to the granular computing mechanism. Within this mechanism, the data is inserted into the particular granule computing procedure to produce particulars. In turn, the particulars generated are injected into the preconcept granule computing procedure to output concepts. Likewise, these preconcepts are sent to the concept granule computing procedure to create concepts. This process gives rise to a context model that is later used by the tacit context reasoning module to assemble decisions that are transferred to the acting component to display actions.

Now, let's illustrate how the tacit context modeling technique would improve the capability of a system to construct a model that represents its environment. The context-aware system would collect data from the environment. Then, the system would identify relations between the data and create the particulars, preconcepts, and concepts. As not all information is relevant to the user, we would have to find a way to discriminate between granules. We suggest giving an activation value to activate or deactivate each kind of granule, which would indicate how active or important is a specific granule. The activation value would be given by three weighted factors, specific to each kind of granule. In this sense, the factors for the activation value for

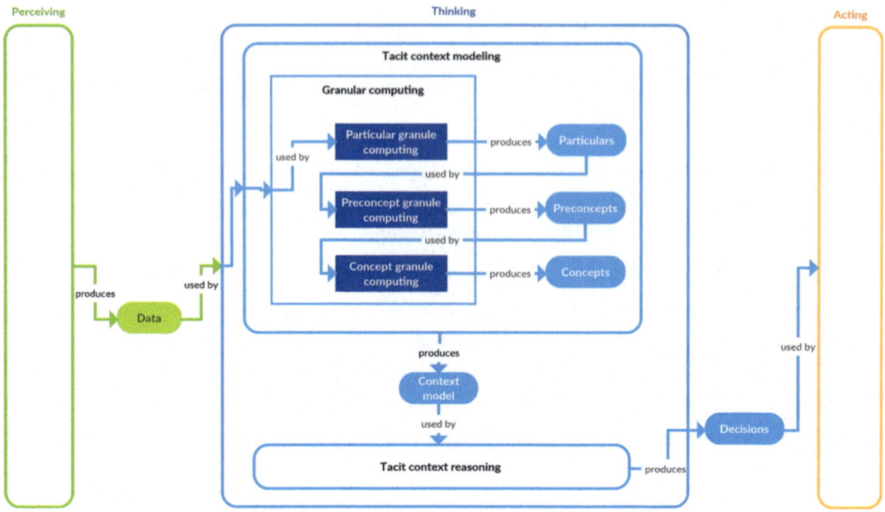

Fig. 12 Tacit context modeling process

particulars would be (1) how much data is associated to it; (2) how many times it is used to create preconcepts; (3) the sum of the activation values of each preconcept it belongs to. The activation value for the preconcept granule would be given by (1) how many particulars it involves; (2) how many concepts it helps to construct; (3) the sum of the activation values of the concepts it is a member of. Finally, the activation value for concepts would be erected from by (1) how many preconcepts it encompasses; (2) in how many decisions it is implicated; (3) the sum of the frequencies of each decision it is entangled in. As this method would enable the context-aware systems to tell apart information, deactivating the information that is not meaningful for the user; and activating and even giving higher activation values to most significant information; we believe that this is a form of constructing the context that best mimics how the brain constructs a model of reality.

5 Conclusions

Ubicomp's ultimate goal is to create technological devices that allow for seamless integration into everyday life. This transparency is only attainable to the extent to which we endow systems with context-awareness. In turn, context-awareness is grounded on the thought that humans effectively communicate ideas; hence, we should try to imitate the process humans carry out to create a model of the world in which they unwind.

Subsequently, we examine the elements involved in the construction of a context-aware device that embraces this invisibleness, and we reached to the conclusion that we need to target the thinking component of the system since this is the one with the analogous function to the human's thinking process. We stated that to better understand this process, we should analyze the knowledge concept. From this, we presented the idea that to improve context-awareness we have to explore the concept of tacit knowledge and analyze the way humans create such knowledge.

In this way, we picked up the tacit knowledge creation process and proceeded to sketch the tacit context modeling technique based on granular computing, which we used to outline the particular granule computing, the preconcept granule computing, and the concept granule computing. We later provide a brief example of how these three kinds of granules can be implemented to construct a context model.

We consider that, by forming these three levels of granules, it will be easier to simulate the process the brain undergoes when constructing its context model, which will allow achieving a model more closely attached to reality. But it will also pave the way to a tacit context reasoning technique that best mirrors the human reasoning process, providing it with a more accurate understanding of the world that surrounds it. This will definitely represent a big step towards improving the decision-making process for the acting component of a context-aware system.

5.1 Future Work

There is still a lot of work to be done regarding this proposal. To get started, experiments must be performed in order to validate and determine the value of this new technique, as well as to make adjustment seeking for the optimal performance. Then, experiments implementing other context modeling techniques need to be carried out with the purpose of comparing results.

The next step is to design the tacit context reasoning, in such a way that it takes advantage of the context model the tacit context modeling technique outputs, but at the same time is grounded on the tacit knowledge creation process. Later, we need to test the performance of the tacit context reasoning technique and compare it to other techniques so that we can determine its value.

Afterwards, it is desirable to examine the efficiency of the thinking component when it is implemented with techniques founded on perceptions of how the brain creates and manages knowledge, the tacit context modeling and the tacit context reasoning.

References

1. Abbagnano, N.: Diccionario de Filosofía. Fondo de Cultura Económica, México (2004)
2. Abowd, G.D., Dey, A.K., Brown, P.J., Davies, N., Smith, M., Steggles, P.: Towards a better understanding of context and context-awareness. In: International Symposium on Handheld and Ubiquitous Computing. Springer, pp. 304–307 (1999)

3. Addis, M.: Tacit and explicit knowledge in construction management. Constr. Manag. Econ. **34**(7–8), 439–445 (2016)
4. Afifi, T.D., Zamanzadeh, N., Harrison, K., Callejas, M.A.: Wired: the impact of media and technology use on stress (cortisol) and inflammation (interleukin il-6) in fast paced families. Comput. Hum. Behav. **81**, 265–273 (2018)
5. Aguilar, J., Jerez, M., Rodríguez, T.: Cameonto: context awareness meta ontology modeling. Appl. Comput. Inform. **14**(2), 202–213 (2018)
6. Alavi, M., Leidner, D.E.: Knowledge management and knowledge management systems: conceptual foundations and research issues. MIS Q. 107–136 (2001)
7. Alegre, U., Augusto, J.C., Clark, T.: Engineering context-aware systems and applications: a survey. J. Syst. Softw. **117**, 55–83 (2016)
8. Augusto, J.C., Callaghan, V., Cook, D., Kameas, A., Satoh, I.: Intelligent environments: a manifesto. Hum. Centric Comput. Inf. Sci. **3**(1), 12 (2013)
9. Bargiela, A., Pedrycz, W.: Granular computing. In: Handbook on Computational Intelligence: Volume 1: Fuzzy Logic, Systems, Artificial Neural Networks, and Learning Systems. World Scientific, pp. 43–66 (2016)
10. Baumard, P.: Tacit Knowledge in Organizations. Sage (1999)
11. Berg-Beckhoff, G., Nielsen, G., Ladekjær Larsen, E.: Use of information communication technology and stress, burnout, and mental health in older, middle-aged, and younger workers-results from a systematic review. Int. J. Occup. Environ. Health **23**(2), 160–171 (2017)
12. Börger, E., Raschke, A.: Modeling Context Awareness, pp. 135–180. Springer, Berlin, Heidelberg (2018)
13. Borges, R.: Tacit knowledge sharing between it workers: the role of organizational culture, personality, and social environment. Manag. Res. Rev. **36**(1), 89–108 (2012)
14. Brooks, S.: Does personal social media usage affect efficiency and well-being? Comput. Hum. Behav. **46**, 26–37 (2015)
15. Brumitt, B., Meyers, B., Krumm, J., Kern, A., Shafer, S.: Easyliving: technologies for intelligent environments. In: International Symposium on Handheld and Ubiquitous Computing, Springer, pp. 12–29 (2000)
16. Cao, Q., Niu, X.: Integrating context-awareness and utaut to explain alipay user adoption. Int. J. Ind. Ergon. **69**, 9–13 (2019)
17. Cao, X., Sun, J.: Exploring the effect of overload on the discontinuous intention of social media users: an sor perspective. Comput. Hum. Behav. **81**, 10–18 (2018)
18. Carvalho, R., Andrade, R., Oliveira, K., Kolski, C.: Catalog of invisibility requirements for UbiComp and IoT applications. In: 2018 IEEE 26th International Requirements Engineering Conference (RE). IEEE, pp. 88–99 (2018)
19. Carvalho, R.M., de Castro Andrade, R.M., de Oliveira, K.M., de Sousa, Santos I., Bezerra, C.I.M.: Quality characteristics and measures for human-computer interaction evaluation in ubiquitous systems. Softw. Qual. J. **25**(3), 743–795 (2017). https://doi.org/10.1007/s11219-016-9320-z
20. Carvalho, R.M., Andrade, R.M., de Oliveira, K.M.: Towards a catalog of conflicts for HCI quality characteristics in ubicomp and IoT applications: process and first results. In: 2018 12th International Conference on Research Challenges in Information Science (RCIS). IEEE, pp. 1–6 (2018)
21. Cheetham, C.E., Barnes, S.J., Albieri, G., Knott, G.W., Finnerty, G.T.: Pansynaptic enlargement at adult cortical connections strengthened by experience. Cereb. Cortex **24**(2), 521–531 (2012)
22. Cheng, X., Yang, J., Xia, L.: A service-oriented context-awareness reasoning framework and its implementation. Electron. Libr. **36**(6), 1114–1134 (2018)
23. Collins, H.: Tacit and Explicit Knowledge. University of Chicago Press (2010)
24. Couderc, P., Kermarrec, A.M.: Enabling context-awareness from network-level location tracking. In: International Symposium on Handheld and Ubiquitous Computing. Springer, pp. 67–73 (1999)
25. Cowan, R., David, P.A., Foray, D.: The explicit economics of knowledge codification and tacitness. Ind. Corp. Change **9**(2), 211–253 (2000)

26. Davenport, T.H., Prusak, L., et al.: Working Knowledge: How Organizations Manage What They Know. Harvard Business Press (1998)
27. Dey, A.K.: Understanding and using context. Pers. Ubiquitous Comput. **5**(1), 4–7 (2001)
28. Dhir, A., Yossatorn, Y., Kaur, P., Chen, S.: Online social media fatigue and psychological wellbeing—a study of compulsive use, fear of missing out, fatigue, anxiety and depression. Int. J. Inf. Manag. **40**, 141–152 (2018)
29. Dobson, S., Denazis, S., Fernández, A., Gaïti, D., Gelenbe, E., Massacci, F., Nixon, P., Saffre, F., Schmidt, N., Zambonelli, F.: A survey of autonomic communications. ACM Trans. Auton. Adapt. Syst. **1**(2), 223–259 (2006). http://doi.acm.org/10.1145/1186778.1186782
30. Eraut, M.: Non-formal learning and tacit knowledge in professional work. Br. J. Educ. Psychol. **70**(1), 113–136 (2000)
31. Forman, G.H., Zahorjan, J.: The challenges of mobile computing. Computer **27**(4), 38–47 (1994)
32. Gaudioso, F., Turel, O., Galimberti, C.: The mediating roles of strain facets and coping strategies in translating techno-stressors into adverse job outcomes. Comput. Hum. Behav. **69**, 189–196 (2017)
33. Grubert, J., Langlotz, T., Zollmann, S., Regenbrecht, H.: Towards pervasive augmented reality: context-awareness in augmented reality. IEEE Trans. Vis. Comput. Graph. **23**(6), 1706–1724 (2017)
34. Haldin-Herrgard, T.: Difficulties in diffusion of tacit knowledge in organizations. J. Intell. Cap. **1**(4), 357–365 (2000)
35. Harris, K.J., Harris, R.B., Carlson, J.R., Carlson, D.S.: Resource loss from technology overload and its impact on work-family conflict: can leaders help? Comput. Hum. Behav. **50**, 411–417 (2015)
36. Hofer, T., Schwinger, W., Pichler, M., Leonhartsberger, G., Altmann, J., Retschitzegger, W.: Context-awareness on mobile devices-the hydrogen approach. In: Proceedings of the 36th Annual Hawaii International Conference on System Sciences, 2003. IEEE, p. 10 (2003)
37. Johannessen, J.A., Olaisen, J., Olsen, B.: Mismanagement of tacit knowledge: the importance of tacit knowledge, the danger of information technology, and what to do about it. Int. J. Inf. Manag. **21**(1), 3–20 (2001)
38. Karvonen, H., Kujala, T.: Designing and evaluating ubicomp characteristics of intelligent in-car systems. Adv. Ergon. Des. Usability Spec. Popul. Part I **16**, 27 (2014)
39. Khabou, N., Rodriguez, I.B., Jmaiel, M.: An overview of a novel analysis approach for enhancing context awareness in smart environments. Inf. Softw. Technol. (2018)
40. Krumm, J.: Ubiquitous Computing Fundamentals. Chapman and Hall/CRC, Washington (2016)
41. Lam, A.: Tacit knowledge, organizational learning and societal institutions: an integrated framework. Org. Stud. **21**(3), 487–513 (2000)
42. Lam, A., et al.: Tacit knowledge, organisational learning and innovation: a societal perspective, vol. 98. DRUID (1998)
43. Lee, A.R., Son, S.M., Kim, K.K.: Information and communication technology overload and social networking service fatigue: a stress perspective. Comput. Hum. Behav. **55**, 51–61 (2016a)
44. Lee, J.H., Hashimoto, H.: Intelligent space–concept and contents. Adv. Robot. **16**(3), 265–280 (2002)
45. Lee, J.W., Helal, S.: Context awareness computing in smart spaces using stochastic analysis of sensor data. In: International Conference on Intelligent Human Systems Integration. Springer, pp. 3–9 (2019)
46. Lee, S.B., Lee, S.C., Suh, Y.H.: Technostress from mobile communication and its impact on quality of life and productivity. Total Qual Manag Bus Excell **27**(7–8), 775–790 (2016b)
47. Lin, H., Yan, Z., Fu, Y.: Adaptive security-related data collection with context awareness. J. Netw. Comput. Appl. **126**, 88–103 (2019)
48. Liu, C., Park, E.M., Jiang, F.: Examining effects of context-awareness on ambient intelligence of logistics service quality: user awareness compatibility as a moderator. J. Ambient Intell. Humaniz. Comput. (2018). https://doi.org/10.1007/s12652-018-1004-z

49. Machado, R., Rosa, F., Almeida, R., Primo, T., Pilla, M., Pernas, A., Yamin, A.: A hybrid architecture to enrich context awareness through data correlation. In: Proceedings of the 33rd Annual ACM Symposium on Applied Computing, SAC '18, pp. 1451–1453. ACM, New York, NY, USA (2018). https://doi.org/10.1145/3167132.3167405
50. Maier, C., Laumer, S., Weinert, C., Weitzel, T.: The effects of technostress and switching stress on discontinued use of social networking services: a study of facebook use. Inf. Syst. J. **25**(3), 275–308 (2015)
51. Mak, B., Nickerson, R., Sim, J.: Mobile technology dependence and mobile technostress. Int. J. Innov. Technol. Manag. (2011)
52. Maran, V., Medeiros Machado, G., Machado, A., Augustin, I., de Oliveira, Palazzo, M.J.: Upcad: a methodology of integration between ontology-based context-awareness modeling and relational domain data. Information **9**(2), 30 (2018)
53. Maravilhas, S., Martins, J.: Strategic knowledge management a digital environment: tacit and explicit knowledge in fab labs. J. Bus. Res. **94**, 353–359 (2019)
54. Meyer, H., Sancho, J.C., Quiroga, J.V., Zyulkyarov, F., Roca, D., Nemirovsky, M.: Disaggregated computing. an evaluation of current trends for datacentres. Procedia Comput. Sci. **108**, 685–694 (2017)
55. Mylonas, P.: The evolution of contextual information processing in informatics. Information **9**(3) (2018). https://doi.org/10.3390/info9030047. http://www.mdpi.com/2078-2489/9/3/47
56. Nonaka, I.: A dynamic theory of organizational knowledge creation. Organ. Sci. **5**(1), 14–37 (1994)
57. Nonaka, I., Takeuchi, H.: The knowledge-creating company: how Japanese companies create the dynamics of innovation. Oxford University Press (1995)
58. Nonaka, I., Toyama, R.: The Knowledge-Creating Theory Revisited: Knowledge Creation as a Synthesizing Process, pp. 95–110. Palgrave Macmillan, London, UK (2015)
59. Nonaka, I., Byosiere, P., Borucki, C.C., Konno, N.: Organizational knowledge creation theory: a first comprehensive test. Int. Bus. Rev. **3**(4), 337–351 (1994)
60. Nonaka, I., Kodama, M., Hirose, A., Kohlbacher, F.: Dynamic fractal organizations for promoting knowledge-based transformation—a new paradigm for organizational theory. Eur. Manag. J. **32**(1), 137–146 (2014)
61. Not, E., Petrelli, D.: Blending customisation, context-awareness and adaptivity for personalised tangible interaction in cultural heritage. Int. J. Hum. Comput. Stud. **114**, 3–19 (2018)
62. Ocegueda-Miramontes, V., Rodríguez-Díaz, A., Castro, J.R., Sanchez, M.A., Mendoza, O.: On modeling tacit knowledge for intelligent systems. Computer Science and Engineering—Theory and Applications, pp. 69–87. Springer (2018)
63. Papaioannou, A.D., Nejabati, R., Simeonidou, D.: The benefits of a disaggregated data centre: a resource allocation approach. In: Proceedings of GLOBECOM, pp. 1–7 (2016)
64. Park, C., Vertinsky, I., Becerra, M.: Transfers of tacit vs. explicit knowledge and performance in international joint ventures: the role of age. Int. Bus. Rev. **24**(1):89–101
65. Pascoe, J., Ryan, N., Morse, D.: Issues in developing context-aware computing. In: International Symposium on Handheld and Ubiquitous Computing, pp. 208–221. Springer (1999)
66. Pedrycz, W.: The principle of justifiable granularity and an optimization of information granularity allocation as fundamentals of granular computing. J. Inf. Process. Syst. **7**(3), 397–412 (2011)
67. Pedrycz, W., Homenda, W.: Building the fundamentals of granular computing: a principle of justifiable granularity. Appl. Soft Comput. **13**(10), 4209–4218 (2013)
68. Pedrycz, W., Song, M.: Granular fuzzy models: a study in knowledge management in fuzzy modeling. Int. J. Approx. Reason. **53**(7), 1061–1079 (2012)
69. Polanyi, M.: The Study of Man (1959)
70. Polanyi, M.: Personal Knowledge—Towards a Post-critical Philosophy. Routledge (1962)
71. Polanyi, M.: The logic of tacit inference. Philosophy **41**(155), 1–18 (1966)
72. Ribeiro, R.: Tacit knowledge management. Phenomenol. Cogn. Sci. **12**(2), 337–366 (2013)
73. Riboni, D.: Special issue on context awareness. Information **9**(3) (2018). https://doi.org/10.3390/info9030067. http://www.mdpi.com/2078-2489/9/3/67

74. Russell, S.J., Norvig, P.: Artificial Intelligence: A Modern Approach. Pearson Education Limited, Malaysia (2016)
75. Ruta, M., Scioscia, F., Gramegna, F., Ieva, S., Di Sciascio, E., De Vera, R.P.: A knowledge fusion approach for context awareness in vehicular networks. IEEE Internet Things J. **5**(4), 2407–2419 (2018)
76. Rutkowski, A.F., Saunders, C.: Emotional and Cognitive Overload: The Dark Side of Information Technology. Routledge (2018)
77. Rutter, M.: Is resilience training the solution to technology overload? Occup. Health Wellbeing **68**(7), 18 (2016)
78. Salo, M., Pirkkalainen, H., Koskelainen, T.: Technostress and social networking services: uncovering strains and their underlying stressors. In: Scandinavian Conference on Information Systems, pp. 41–53. Springer (2017)
79. Sanchez, R.: Tacit Knowledge Versus Explicit Knowledge: Approaches to Knowledge Management Practice, p. 191. TEAM LinG (2005)
80. Satoh, I.: Evolutionary mechanism for disaggregated computing. In: 2012 Sixth International Conference on Complex, Intelligent and Software Intensive Systems (CISIS), pp. 343–350. IEEE (2012)
81. Satyanarayanan, M.: Fundamental challenges in mobile computing. In: Proceedings of the Fifteenth Annual ACM Symposium on Principles of Distributed Computing, pp. 1–7. ACM (1996)
82. Schmidt, A., Aidoo, K.A., Takaluoma, A., Tuomela, U., Van Laerhoven, K., Van de Velde, W.: Advanced interaction in context. In: International Symposium on Handheld and Ubiquitous Computing, pp. 89–101. Springer (1999)
83. Selinger, E., Feist, G., Stone, D.: Erratum to: tacit knowledge: new theories and practices. Phenomenol. Cogn. Sci. **2**(12), 251–251 (2013)
84. Sonnentag, S.: Being permanently online and being permanently connected at work: a demands-resources perspective. Permanently Online, Permanently Connected, pp. 258–267. Routledge (2017)
85. Spínola, R.O., Silva, J., Travassos, G.H.: Characterizing ubicomp software projects through a checklist. In: Proceedings of the I Workshop on Pervasive and Ubiquitous Computing, Gramado, Brasil (2007)
86. Spínola, R.O., Pinto, F.C. do R., Travassos, G.H.: Ubicheck: an approach to support requirements definition in the ubicomp domain. In: Proceedings of the 2010 ACM Symposium on Applied Computing, pp. 306–310. ACM (2010)
87. Stenmark, D.: Leveraging tacit organizational knowledge. J. Manag. Inf. Syst. **17**(3), 9–24 (2000)
88. Sultan, N.: Reflective thoughts on the potential and challenges of wearable technology for healthcare provision and medical education. Int. J. Inf. Manag. **35**(5), 521–526 (2015)
89. Takayama, L.: The motivations of ubiquitous computing: revisiting the ideas behind and beyond the prototypes. Pers. Ubiquitous Comput. **21**(3), 557–569 (2017)
90. Temdee, P., Prasad, R.: Context-Aware Communication and Computing: Applications for Smart Environment. Springer (2008)
91. Theng, Y.L.: Ubiquitous Computing: Design, Implementation and Usability: Design. Implementation and Usability. IGI Global (2008)
92. Virtanen, I.: In search for a theoretically firmer epistemological foundation for the relationship between tacit and explicit knowledge. Electron. J. Knowl. Manag. **11**(2), 118 (2013)
93. Weiser, M.: The computer for the 21st century. Sci. Am. **265**(3), 94–105 (1991)
94. Weiser, M.: The future of ubiquitous computing on campus. Commun. ACM **41**(1), 41–42 (1998)
95. Weiser, M., Gold, R., Brown, J.S.: The origins of ubiquitous computing research at parc in the late 1980s. IBM Syst. J. **38**(4), 693–696 (1999)
96. West, M.T.: Ubiquitous computing. In: Proceedings of the 39th Annual ACM SIGUCCS Conference on User Services, pp. 175–182. ACM (2011)

97. Yürür, Ö., Liu, C.H., Sheng, Z., Leung, V.C., Moreno, W., Leung, K.K.: Context-awareness for mobile sensing: a survey and future directions. IEEE Commun. Surv. Tutor. **18**(1), 68–93 (2016)
98. Zheng, Y., Shekhar, S., Jose, A.A., Rai, S.K.: Integrating context-awareness and multi-criteria decision making in educational learning. In: Proceedings of the 34th ACM Symposium on Applied Computing (2019)

Applied Decision Making in Design Innovation Management

Eduardo Ahumada-Tello, Manuel Castañón-Puga, Carelia Gaxiola-Pacheco and Richard D. Evans

Abstract This article focuses on decision making for innovation management during product design. As such, it requires a holistic approach that determines the elements necessary for decision-making in knowledge-based organizations. The first part of this chapter explores the importance of the decision-making process, its role in organizational growth and development of strategies, as well as the high complexity that these processes have acquired over time in global and turbulent markets. Then, the decision-making process is detailed, recounting the elements that form it and the impact it has on organizations. Innovation and its classifications are later described, as well as its importance to modern economies. The importance of design, management and product development is also described. Case studies of companies operating in the Information Technology sector in Baja California, Mexico, are used to contextualize the research and demonstrate its validity for use during the development of new software products. Finally, the chapter concludes by exploring the relevance of this study, from a complex approach, using computational methodologies that increase the feasibility of carrying out decision-making with greater sustenance and expectation of success.

E. Ahumada-Tello (✉)
Facultad de Contaduría y Administración, Universidad Autónoma de Baja California, Calzada Universidad 14418, Mesa de Otay, 22390 Tijuana, B.C., Mexico
e-mail: Eduardo.Ahumada@uabc.edu.mx

M. Castañón-Puga · C. Gaxiola-Pacheco
Facultad de Ciencias Químicas e Ingeniería, Universidad Autónoma de Baja California, Calzada Universidad 14418, Mesa de Otay, 22390 Tijuana, B.C., Mexico
e-mail: puga@uabc.edu.mx

C. Gaxiola-Pacheco
e-mail: cgaxiola@uabc.edu.mx

R. D. Evans
Department of Design, College of Engineering, Design and Physical Sciences, Brunel University London, Kingston Lane, UB8 3PH Uxbridge, UK
e-mail: Richard.Evans@brunel.ac.uk

M. A. Sanchez et al. (eds.), *Applied Decision-Making*, Studies in Systems, Decision and Control 209, https://doi.org/10.1007/978-3-030-17985-4_5

1 Introduction

The decision-making process is fundamental to the achievement of organizational growth. It is also a key element for strategic actions that take place in organizations. This process encapsulates a series of characteristics that deem it a system, with a high level of complexity, involving multiple organizational actions, which require human, technological and informational elements.

The human element involved in the decision-making process is responsible for making correct choices that guide the organization towards its short, medium and long-term objectives [1], starting with its position in the market; this is where it competes, experiences growth and, in future, obtains competitive advantages that give it the possibility of increasing its number of clients, its market share and the loyalty of existing customers.

On the other hand, developments in technology have given way to an exponential increase in information availability, which allow organizations to generate the best solution alternatives, based on this data. Data warehouses, data mining, machine learning, business intelligence and data science are some of the technological approaches to managing the information available to the decision maker to obtain sufficient elements that ensure maximum success in the solution of any problem within an organizational context.

When the decision-making process involves the design of new products, it must be complemented by external and internal information that facilitates the projection of the expected results before the growth of the organization. Turbulent business environments create opportunities for the development of new products and, at the same time, increase the number of challenges that organizations have to overcome to deliver innovations in their product line or in the development and modification of existing products.

The capabilities of organizations allow the generation of new products and processes to accommodate these actions according to market conditions; this is based on the theory of resources that supports the development of innovation within the company and helps predict the behavior of the market in which the products are sold [2, 3].

2 The Decision-Making Process

Decision-making is an intentional process that combines the analysis of information, use of technology, evaluation of alternatives, and the decision itself. Strategic decision-making is crucial for implementing business strategies. The decision-making process is carried out by identifying the best alternative to solve a problem in the organizational environment. This is achieved through the collection of information and the evaluation of all possible solutions to the current situation or phenomenon. For this reason, the process for considering the best option can be determined through a strategy of steps that help to take deliberate and thoughtful actions using the information at hand and thereby increase chances of success. By using this approach, it increases the chances that the chosen option is the most satisfactory and provides the best possible chance of achieving success.

In Fig. 1, the relative steps in the decision-making process are shown. The first action to be taken is to accept that a decision is required. The next step is to compile certain pertinent information before making a decision; this must include relevant and reliable data. In order to achieve this second step, internal and external information must be used by the organization; hence, the selection of information sources is an important step for the success of this step.

The following step is to identify possible alternatives. This is achieved through the collection of information i.e. once the volume of available data increases, options or possible paths of action will emerge. During this step, creativity and imagination play a crucial role—the more concepts applied, the more options and action plans can be integrated into the analysis of the final solution.

In the fourth step, the evidence located during the information search must be evaluated to predict results that can be obtained by applying each of the alternatives. It is necessary to validate that the identified need, found in the first step, can be solved by using each alternative. In this step, all possible and desirable alternatives are listed. As the organization goes through this internal process, certain alternatives that appear to offer a greater potential to reach their goal are seen. Finally, the alternatives are placed in an order of priority, based on the previous evaluation. In the next step, one of the options is chosen. Once evaluation has been completed, the organization proceeds to the selection of the most suitable solution to solve the situation. Then, they proceed to execute the actions required to implement the selected alternative

Fig. 1 General steps in the decision-making process

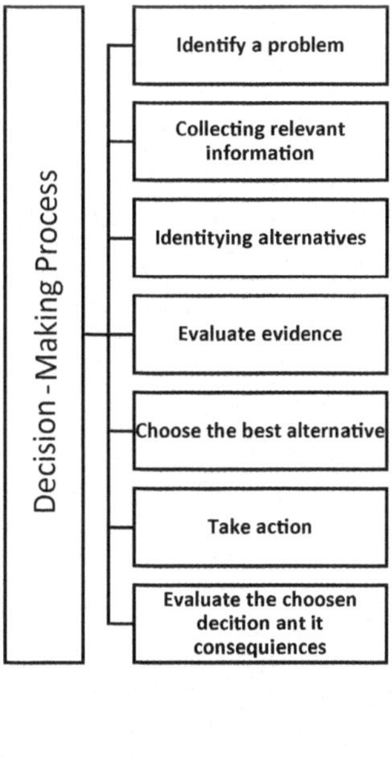

and, finally, they carry out a process of review and evaluation of the decision and the consequences that arose before the decision.

Some authors consider the decision-making process to be influenced by activities that are internal to the organization, such as knowledge management, leadership style and organizational performance [4]. In this study, a regulatory framework is proposed that supports the relationship between the factors that support knowledge management, such as collaboration among members of the organization, skills in information technologies and the mediating process of knowledge, which increases during the design, and the development and implementation of new products.

One method of decision making is through the creation of decision trees. This works by extending decision nodes, opportunity nodes and result nodes. Figure 2 shows the possible alternatives and the possible slopes that are obtained by developing different edges of the nodes that lead to decision making. In this way, the following decision paths are observed:

Table 1 shows that there are seven possible combinations of actions indicated in the decision tree and, therefore, seven possible outputs to be evaluated to determine the best alternative solution. In this case, a success percentage is indicated in the route and has been included since this is an important factor when performing the evaluation and forecasting the process of each of the options; this data can be decisive in selecting which route will continue to meet organizational objectives and, therefore, which is the one that will meet expectations.

Finally, the managers of the organization have the responsibility to lead the organization towards economic growth and commercial success [5]. In this case, strategies are observed that are oriented towards compliance with organizational objectives. It is in this sense that the versatility of knowledge is formed from the information obtained during the decision-making process. This includes opinions of experts and the innovation that emerges when determining the course of action to be taken. The process of knowledge generation is a dynamic, multi-dimensional and complex pro-

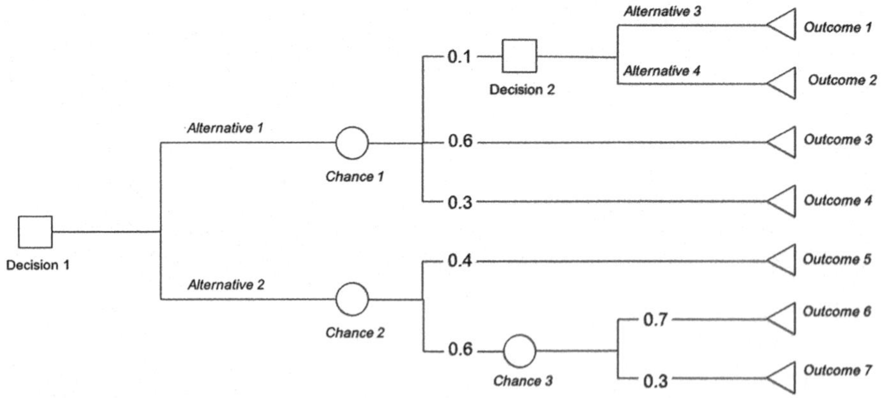

Fig. 2 Example decision tree [19]

Table 1 Decision tree route outcomes

Route	Graph traceability
A	D_1 ➔ A_1 ➔ -Ch_1 ➔ Success 10% ➔ D_2 ➔ A_3 ➔ Outcome 1
B	D_1 ➔ A_1 ➔ -Ch_1 ➔ Success 10% ➔ D_2 ➔ A_4 ➔ Outcome 2
C	D_1 ➔ A_1 ➔ -Ch_1 ➔ Success 60% ➔ Outcome 3
D	D_1 ➔ A_1 ➔ -Ch_1 ➔ Success 30% ➔ Outcome 4
E	D_1 ➔ A_2 ➔ -Ch_2 ➔ Success 40% ➔ Outcome 5
F	D_1 ➔ A_2 ➔ -Ch_2 ➔ Success 60% ➔ Ch_3 ➔ Success 70% ➔ Outcome 6
G	D_1 ➔ A_2 ➔ -Ch_2 ➔ Success 60% ➔ Ch_3 ➔ Success 30% ➔ Outcome 7

cess that directly impacts on the innovative journey of the organization. It is from this new knowledge that strategies and action pathways emerge which will position the company as an agent of influence.

3 Defining Innovation

In business and social environments, innovation has been defined in many ways. It is a novel concept and, therefore, there are different views on what is the best way to refer to this construct [6]. In this chapter, we will refer to the concept of innovation as "the act of introducing a new device, method or material with a commercial application or with practical objectives" [7]. Therefore, it is not enough just to develop a product, service or process; it is also necessary to apply it in the commercial field for this concept to be fulfilled. Sometimes, the development of new products and the innovation process is part of new and serious challenges in the business environment, where the organization needs to stay competitive; in other cases, it is because of the entrepreneurial spirit of the leaders in charge of the organization [8].

It is recognized that there are different classifications of innovation, including product innovation, process innovation and service innovation. At the same time, these differ in terms of incremental innovation and radical innovation. Finally, modular and architectural innovation are mentioned throughout the literature. In the general description of each of the concepts or classifications, the purpose and objectives to be achieved must be considered to implement that type of innovation, which is more akin to organizational expectations. From certain points of view, innovation can be motivated by the buying behavior of customers, while there is even a classification of different types of clients, from early adopters of a certain innovation or new product to those who do not change or buy new products until they have no option, but to purchase [9].

a. **Product, Service and Process Innovation**

Product innovation relates to items that are tangible. In contrast to these physical products, services are generally not tangible. To understand the differences between physical products and non-tangible services, an example is of a supermarket which delivers groceries to the door of the customer's home. Carrying the groceries to the customer's home is a new service for the supermarket, resulting in service innovation. Innovation of products and innovation of services have commonalities that affect the outside of the organization, in terms of what they produce.

Process innovation relates to changes in how companies conduct their activities. Such an example is where, in a company, the process of data collection is carried out manually by employees (i.e. by means of entering data, typically in paper format) and then reporting it during meetings but, over time, this is now achieved through electronic means that allow the data to be analyzed. This process may be seen to have been innovated i.e. process innovation.

Innovation in products refers to improvements that are made to the products being developed by an organization. A relevant example of what product innovation involves is the improvement and updating of an existing product offered by a company or the development of a new product. In this case, telephones provide a relevant example of a product that has experienced constant improvement, while also providing other types of customer satisfactions, such as music, access to climate information, geographical data and more. In this case, the customer pays for the product (the telephone) and services (communication), but also benefits from additional intangible products that provide satisfaction and meet end-user needs.

The creation of the bar code can also be provided as an example. Previously, employees of a store or supermarket learned by heart the prices of available products. Now, each product has been provided with a unique number in bar nomenclature, which has made it easy to scan. This change in how to pay for the products is classified as process innovation. Likewise, we have the example of the assembly line to manufacture cars which Henry Ford developed. In this way, the car advances through the facilities while it is being assembled by specialized personnel until it finally ends with full functionality. This form of industrial work leads to more efficient and higher quality production processes. This innovation also relates to a process.

b. **Incremental and Radical Innovation**

As described, innovation can be focused on products, services or processes. These three types of innovation can be incremental or radical. Incremental innovation refers to adjustments, improvements and innovations to solutions that already exist and are in the market. Radical innovation typically relates to new solutions or major changes to existing products.

An example of incremental innovation is the improvement in the presentation of products. For example, the food industry often produces changes to the appearance of its products to meet the demands of younger consumers. For this reason, packaging is often redesigned, with organizations adjusting colors, logos, fonts, and other characteristics that keep the product in constant demand; this is considered incremental as the changes are minor and fit the goal of maintaining sales.

On the other hand, an example of radical innovation is the emergence of a new solution, which may be a product, a service or a process. In the same way, we can take the example of mobile telephony. In this industry, communication protocols, such as 3G, are used which, when implemented, caused a relevant change in the industry. The 3G network increased the possibilities of developing applications that are more demanding for telephone devices—for example, screens are now able to change color to take advantage of new capabilities. The design of new telephones has also evolved, with work activities now being able to be performed through this means of communication. The same happened when the 4G network emerged; in both cases, these changes are so relevant that they directly impact the entire sector.

c. **Architectural and Modular Innovation**

Modular innovation refers to changes that can be made to a component without changing the configuration of the system itself. This can be understood as improvements to components that result in innovations to that specific part, but the system or solution continues to function in its original manner. Innovation of the architecture refers to a change in the configuration of the complete system and how its parts or components now interact. Due to the nature of these innovations, they are applicable to relatively complex and technical systems.

An example that illustrates the difference between architectural innovation and component innovation can be the camera. Previously, professional cameras used photographic rolls to capture images. These cameras could be improved operationally by exchanging lenses, lights, protective devices etc. but, ultimately, the camera and its components continued to interact in the same way. These components, used to improve the performance of the camera, were part of an incremental innovation. However, when the camera changed to become digital, then the components did not interact in the same way. For example, communication interfaces, storage media, types of lenses and all other attachments were modified, and their operation altered. This is, therefore, architectural innovation.

This distinction in types of innovation is important due to the operations of companies that are involved in the development of the components of each system. In the case of carrying out a radical change leading to an architecture innovation, it is possible that the companies are not ready to manufacture or participate in the changes that arise. This is due to the fact that companies are usually operating around the product generated and usually have assembly or production lines specialized in certain functions; for example, following the example of the camera, there is specialization in lenses, storage media, etc. If these changes happen incrementally, it is feasible to make adjustments, but if the change is of an architectural nature, then they may not be able to adjust to this new component; for this reason, technological monitoring is essential in order to avoid this possible scenario.

d. **The Role of Innovation in the Modern Economy**

Innovation is a phenomenon that has permanently existed in the history of man, from the invention of the wheel, writing, the means of agricultural production to modern computers, the internet and the interaction with artificial intelligence. We

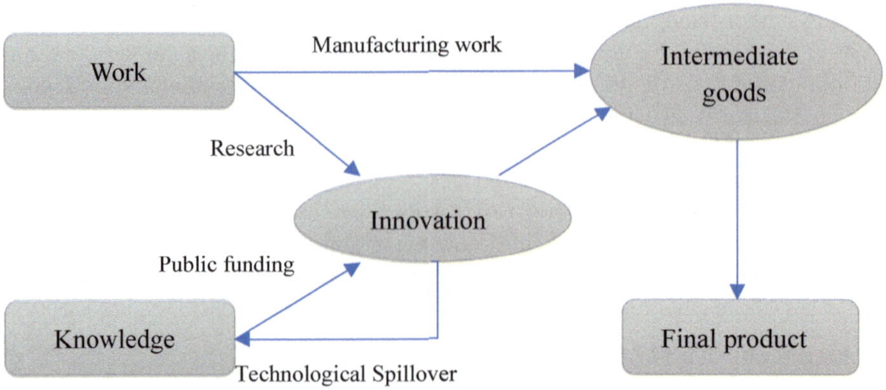

Fig. 3 Innovation in the product development process

have permanent presence of this activity in the lives of people and society. Over time, academics have addressed the issue of innovation and its impact on economic and social development in social groups. Studies have taken into consideration economic, industrial and technological perspectives to those that involve the anthropological and social part of the influence of innovation in society. Regarding economic aspects, we can see in Fig. 3 the role played by innovation in the productive process for the development of new products.

From the approach shown in Fig. 3, it is observed that in order to complete the innovation process, it is necessary to combine work, through research, as well as knowledge, through public funds. These processes impact on technological spillovers and intermediate goods. In the end, the joint process has a direct impact on the final product that is to be developed or that is being innovated for the competitive benefit of the company [10].

Figure 3 shows that the innovation development process is part of the productive processes that contribute ideas and new products to the market. The work of manufacturing and innovation go hand-in-hand to generate intermediate goods that are raised and that give way to the creativity and design of new products for an increasingly demanding market.

4 New Product Development, Management and Design

The New Product Design process and technological innovation process have similar characteristics. The phases that are part of design can be observed in parallel to the programming of innovation in products, services or processes [11]. Based on the initial information gathered, the basic conceptualization of a product or the technological innovation that is intended to be carried out is developed. During the second phase, the prototype is designed, or the new function is developed. Afterwards, the design adapts to its commercialization and the invention becomes an innovation when applied to a product, service or process. Finally, the expected results are evaluated and contrasted with the real ones. The adjustments are made with a satisfactory and common result for both processes, where the objective is to increase firm competitiveness.

During the process of launching new products, a key element is the design contact with the product's target market [12]. Market analysis is carried out to detect existing opportunities and, in turn, the capabilities available to the company to determine the strategies for launching new products based on creativity and the design process, as well as research and development illustrated in Fig. 4. The search for ideas is carried out to clarify the content and characteristics of the new products that focus on the definition of the concept of a new product, possibly by leaders intuition [13], and it then enters into competition with the rivals of the industry to satisfy the needs of the market, as well as the analysis of profitability for the investors involved in the development. In the end, it is decided analytically to approve the product design internally, and externally to the company for its realization, industrialization and through a marketing plan, its introduction to the market.

Fig. 4 New product design

5 Case Study: Information Technology Firms in Baja California

Firms operating in the IT sector of Baja California, Mexico, arose mostly in the 1990s. Its growth has been maintained regularly since the second half of that decade until today. Firms are mainly concentrated in Tijuana and Mexicali, as well as some being situated in Ensenada and Tecate. Although the average level of employment has grown in recent years, it is still, as in the rest of the sector in Mexico, essentially made up of microenterprises with less than 15 employees. The scope of this work has been the municipality of Tijuana, Baja California. The holistic research focus includes a qualitative and quantitative approach. In the first case, interviews were carried out with managers of fourteen IT firms. In the following approach, all interviews were conducted through a questionnaire, which was directed at targeted individuals and personally administered to the key informant. The 14 research subjects were the owners and/ or managers of information technology companies.

a. **Research Problem**

The influence of the decision-making process, the design of new products and the management of innovation in organizational performance of companies in the information technology sector is unknown. In these firms, a study was carried out on the decision-making process, the design and development of new products and innovation management, as determining factors for improved organizational performance reflected in the firm competitiveness. This research process led to obtaining the following results in a regression study performed on the data from a survey applied to managers and owners of these firms in the IT sector.

In order to fulfill the objectives of this research, it was necessary to operationalize the theoretical and conceptual framework. The research objectives and questions, along with the study variables and hypotheses are used as the indicators to build the measurement instrument. In this chapter, it applies to empirical knowledge to gather information from the field. Once this stage is finished, the information must be processed. The following paragraphs detail the following:

- Designing measuring instruments.
- Measuring scales.
- Instrument reliability tests and codification.
- Data collection.
- Data processing.
- Qualitative analysis (coding, network development and semantic construction).
- Quantitative analysis (statistical analysis).

The instruments chosen for data collection were semi-structured interviews via questionnaire. The questionnaire included multiple options from specific scales and provided information for the analysis of each one of the variables that influence the factors that determine the organizational performance in information technologies. In Fig. 5, we illustrate the methodology used.

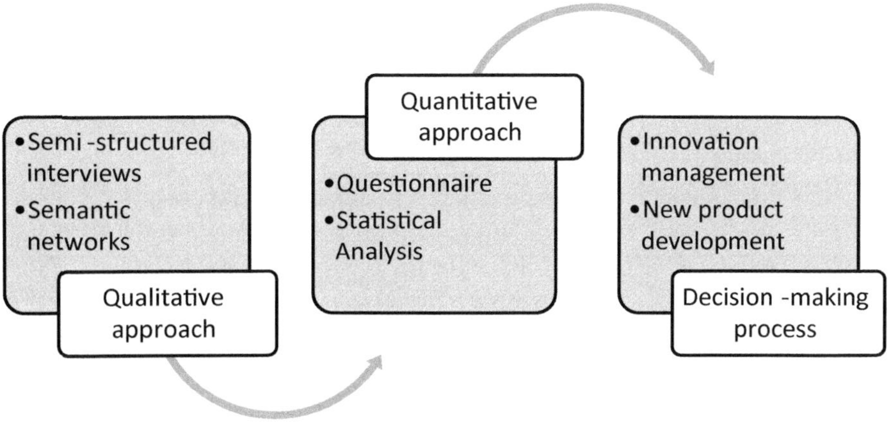

Fig. 5 Methodology adopted

In this case study, we aim to determine how, from a qualitative and quantitative approach, models can be developed for representing the decision-making process in the development of new products and the management of innovation. For this reason, from the application of semi-structured interviews and semantic networks, as well as applying questionnaires and the development of statistical analysis, it is possible to create models that facilitate the decision-making process.

b. **Research Aim**

The aim of this research is to develop a methodology to increase the accuracy or decision-making process, including qualitative and quantitative analysis, with simulation models from statistical data in the design of new product development from management of innovation in organizational performance in companies in the IT sector in Tijuana, B.C.

c. **Qualitative approach to business research**

This first phase of this research consisted of carrying out semi-structured interviews with fourteen managers of medium and large IT companies in Baja California, all of which having direct responsibility in their organization for influencing the economic development of the city of Tijuana.

On this basis, the following technology-based companies present in the Tijuana region, and in the surrounding areas were considered, justifying this extension of the geographical area under the argument that despite some of them not being in the limits of the municipality, they are deemed to influence the competitive development of the city. Table 2 shows the companies that were part of this study.

Table 2 Interview sample

No.	Empresa	Notes
1	Gameloft	Companies in the IT sector dedicated to the construction of videogames and software development Aerospace sector Promoter of business training Services and development of software solutions Automotive industry Information technologies
2	Honneywell Aerospace	
3	CETYS	
4	Grupo Red	
5	Skyworks	
6	Kenworth	
7	Ajosía	
8	Scantibodies	
9	Baja Innova	
10	Softek	
11	CICESE	
12	Technys	
13	Siscor	
14	Laboratorios y Soluciones Genéticas	

d. **Quantitative approach to business research**

i. **Reliability test**

First, we conducted an analysis on the questionnaire developed for this research by checking the following steps to perform and analyze the instrument:

- Content: It is the degree to which the questionnaire reflects a specific domain of the content to be measured.
- Discriminating: It is the degree to which a construct differs from other constructs.
- Construct: It refers to the degree to which a measurement is consistently related to other measurements, according to the theory about the constructs being measured.
- Convergent: Degree in which the same results are obtained in the measurement of a concept when applying different measurement methods.
- External: Degree in which it is ensured that the sample is representative of the population to which it is desired to generalize.

Then, we performed the reliability analysis of the measuring instrument in order to guarantee its performance on the selected subjects.

- Description of the structure of the questionnaire for each independent variable.
- Reliability analysis through the implementation of Cronbach's Alpha to each dimension.
- Final description of the structure of the questionnaire.
- Resulting reliability analysis.

Cronbach's Alpha was acceptable for each variable with the following results: for the variable organizational performance, it achieved a 0.895 value; for the Decision

Making Process (DMP), the variable achieved was 0.932; for New Product Development (NPD), it achieved a 0.911 value; and finally, for the Innovation Management (IM) variable, it achieved a 0.925 value. This means that in all cases, a high internal consistency between the items of each variable was achieved.

ii. **Regression analysis**

Multiple regression is a statistical technique through which the relationships between different variables can be analyzed. For example, between dependent or criterion variables and a group of independent or predictive variables. As a statistical tool, multiple regression is frequently used to achieve three objectives:

1. Find the best predictive equation for a group of variables, for example, given X and Y as predictive variables, which is the value of Z the dependent or criterion variable.
2. To control confusing factors and to find out the influence of a variable on a group of variables. For example, the identification of independent relationships.
3. To find structural relationships and provide explanations for apparently complex multivariate relationships.

For the purpose of this research, the dependent variable is Organizational Performance (OP), and the predictor variables are Decision Making Process (DMP), New Product Development (NPD) and Innovation Management (IM). The results obtained are reflected in Table 3:

The results shown in Table 3 indicate that the most relevant variable to increase organizational OP is the DMP which, as reflected in the regression equation, influences in a highly significant way, by achieving a multiplication factor of 0.403. The variable NPD is the next to influence the dependent variable, with a factor of 0.152, and finally the variable IM with a factor of 0.102. In this way, all the independent variables included in this study influence the OP and improve the conditions of the organizations from the point of view of the subject of study in these technology-based companies and, therefore, it is possible to understand the phenomenon that sustains our research.

The recommendation obtained from the results of this study encourages further improvements in the organization following the statistical analysis of the study. For example, in relation to the DMP, it is the activity that must be professionalized more, following the trends and strategies that are current in the business environment of technology-based companies worldwide. This means that a company must manage its data and information in such a way that improves the elements that lead it to make strategic decisions.

Table 3 Regression model

Model	Regression equation	Method
1	$OP = -4.876 + 0.152\ NPD + 0.102\ IM + 0.403\ DMP$	Forward at 10%

As a next strategy to be implemented, it is established that NPD and IM maintain a similar level of influence for the improvement of PO. This is why both strategies or variables are positioned at a level of importance lower than the DMP. This confirms that the process to reach the decision to develop strategies, implement functions, change activities or any other action that is encouraged in the organization, becomes a high impact factor to improve the organization itself.

e. **Discrete modelling**

Figure 6 illustrates the regression model indicated in the previous section. We used Extend® Simulation tool to replicate the variables behavior, while in Fig. 7, we reflect the output result of this computational model. From this, we observed the model behavior and it can be confirmed and checked in a similar way to the empirical data.

The tests carried out using the model confirm the validity of the regression equation. According to this, it is feasible to represent the outputs of an equation via discrete model, however, this representation emerges from a statistical model and it is recommended to continue with the application of a functional model in a tool for social modeling. An example of the following applications works against the dilemma of using simulation software to represent the interaction between agents and the effect on the study variables.

f. **Modelling development**

The next steps in this research is to propose the analysis by new tools and procedures that help to model a realistic social system and improve the decision-making process according to this. This cannot be achieved by resorting to only one particular type of architecture or methodology [14]. There are several methodologies that we could apply to develop alternate approaches to analyze this phenomenon.

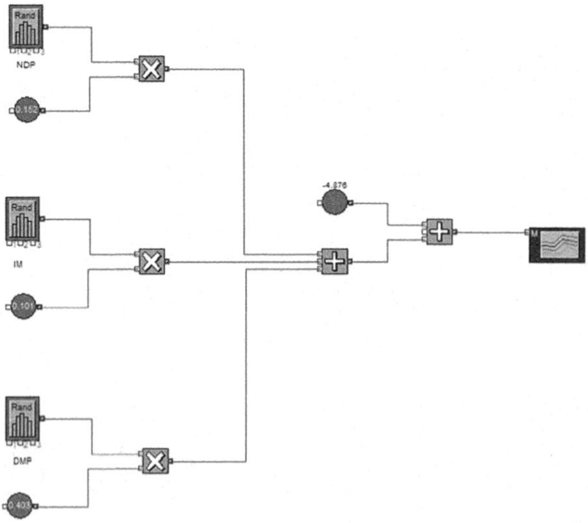

Fig. 6 Model representation in Extend®

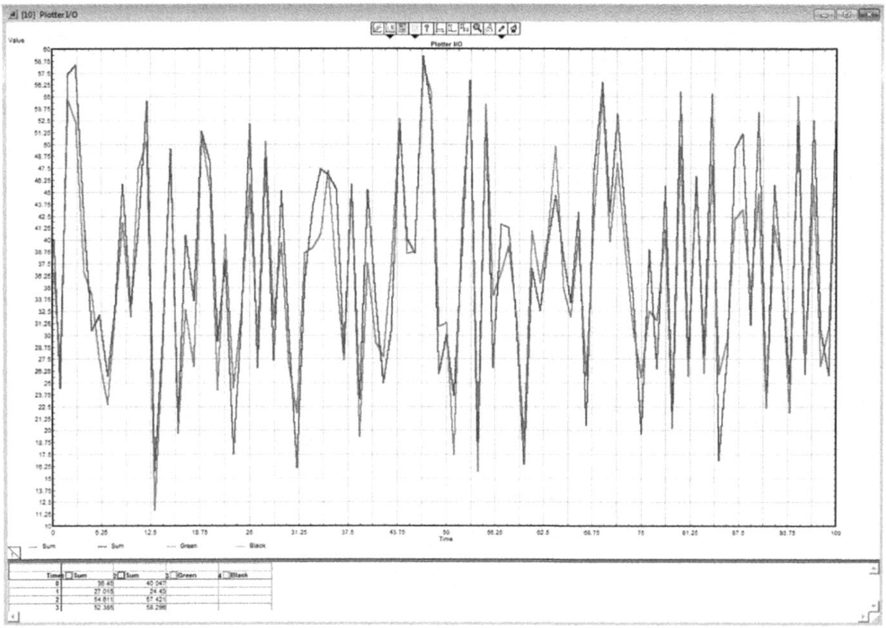

Fig. 7 Model outcome in Extend®

Distributed Agency (DA), Belief-Desire-Intentions (BDI), Fuzzy Logic and others are considered. These methodologies represent potentially ground-breaking applications in social science areas, which minimize the natural distances between physical and sociological systems approaches by presenting several scenarios using data with computational tools and therefore increasing the accuracy of the decision-making process in industrial activities.

In this work, we lay the foundations for a description of socio-economic phenomena, in a process that weaves different available computational techniques to represent behavior in a contextualized business environment, by analyzing agents with limited rationality and complex interactions [15]. These actions represent an approach to simulation architectures, creating a language that links the social sciences to programmable terminology and that can thus be broadly applied. The modelling methodologies represent a general theory of collective behavior and structure formation, which intends to redefine agency and reflect it in multiple layers of information and interaction, as opposed to traditional approach in which agency is only reflected in individual isolated agents [16].

To build the computational model, it is necessary to follow the distributed agency methodological steps listed as follows [17]: 1. Determine the levels of agency and their implicit relationships; 2. Data mine 3. Generate a rule-set; 4. Multi-Agent Modelling (Implementation on an agent-based simulation tool); 5. Validate the model; 6. Conduct a simulation and optimization experiment; and 7. Analyze the outputs.

Although the methodology covers the life-cycle of a research process, we are describing the data mining and generating rule set steps. For the purpose of this chapter, we have focused on the neuro-fuzzy approach in order to set up a rule set into agents. An Interval Type-2 Fuzzy Neural Network (IT2FNN) are used for automatically generate rules. The phase of data mining using Interval Type-2 Fuzzy Logic Systems (IT2FLS) becomes complicated, as there are rules to determine which variables to take into account [16].

Using IT2FNN for automatically generating the necessary rules, this phase of data mining with an IT2FLS becomes complicated, as there are enough rules to determine which variables one should take into account. Using search method of back propagation and hybrid learning with the intention of being more efficient, even with the use of other methods such as genetic algorithms [18]. Since the IT2FNN method seems to produce more accurate models with fewer rules, we could represent and thus illustrate in Fig. 8, where you can observe the path for the selection of the correct output in the rules designated to improve organizational performance.

This optimization algorithm is widely used and is a numerical method to minimize an objective function in a multidimensional space, and the approximate global optimal solution to a problem with N variables, which minimize the function varies smoothly [16]. Using this grouping algorithm, we obtain the rules, the agent receives inputs from its environment and also must choose an action in an autonomous and flexible way to represent, in this case, the decision-making process that affect new product development and innovation management and therefore the impact on organizational performance.

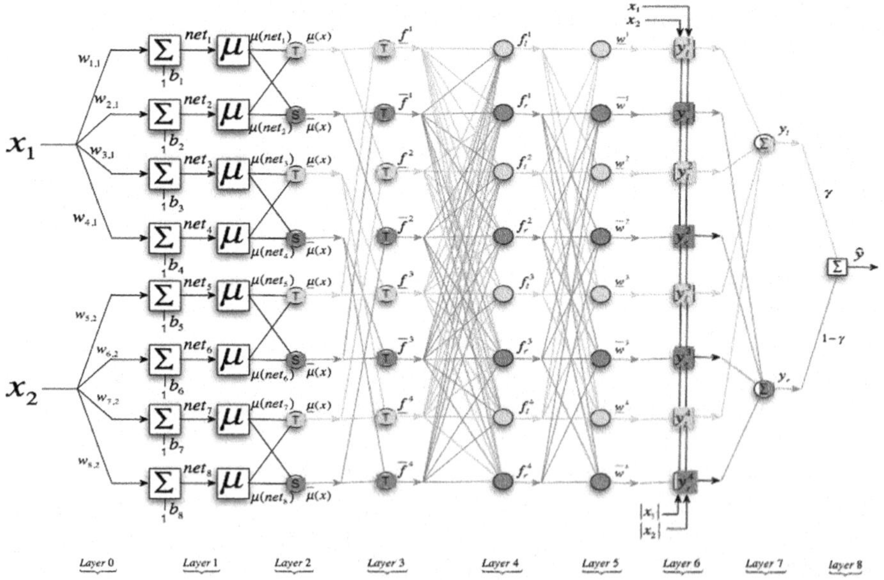

Fig. 8 Architecture IT2FNN applied to analyze social phenomena [20]

6 Conclusions

Management innovation and product design technologies facilitate knowledge creation. In the most competitive, complex, and turbulent environments, knowledge is the most important resource for firms and territories. In such environments, organizations must use their knowledge and managerial skills to improve their competitive position taking advantage of management strategies to promote innovation and product design and development. The choice of qualitative research methodology is due to its function of helping to deeply understand the dynamics present within individual scenarios and to discover new relationships and concepts within them, its application focuses on understanding rather than verification or validation of previous presuppositions. Case study researchers focus on a program, event or activity that involves individuals rather than groups of individuals. There is also an interest in describing the process rather than identifying patterns of behavior exhibited by the group. This is why it is important to model behavior within the organization about decision-making process, which is a multivariable phenomenon.

One of the first qualitative findings is that knowledge is presented as one of the assets that can give companies the greatest value. The study indicates that the leaders of the organizations that were studied find an important reference towards knowledge. From intellectual property and technology transfer to innovation, research and development activities, the key importance of this type of intangible assets for the growth and improvement of organizational performance in technology-based companies located in the Tijuana is very relevant. The importance of the IT sector is reaffirmed when this sector is updated and increases its value, with better use of management innovation processes. People and processes are fundamentally logical-mathematical in nature, this means that many of their activities revolve around the value of knowledge and how it manifests itself in the possibility of better performance for the organization.

Innovation is also considered one of the aspects that help the learning of companies and their strengthening in the market of information technologies. For this business strategy to be successful, it must have a highly specialized, trained and certified workforce. The organizational culture, public policies and the link with the educational sector is fundamental, the main reason for being dependent on these conditions is that it is necessary to carry out a cultural change with a greater focus on quality, sustainable development and in the respect as well as the protection of intellectual property, all of them foundations of an environment of innovation, research and development as an edge of both economic growth and the development of systemic competitiveness in companies in the sector. This is where the importance of management innovation, new product development policies and the adequate implementation of an effective decision-making process come to be the most important business strategies recommended for this firms.

It is in this context that the research is based on the statistical methods that emerge from the quantitative study. It is verified that the three variables intervene directly in a positive way in the increase of the systemic competitiveness. The evaluation of the data obtained empirically with the statistical analysis carried out using SPSS software, confirm a positive relationship of the three predictor variables with the dependent variable.

This means that the organizational performance increases as a result of the increase in strategies for improve decision-making process, management of innovation and development of new products. These three factors are confirmed in the results of the linear regression statistical study and are explained in the discrete model that has been carried out using the Extend simulation software. From these results, we proceed to the development of a model based on the theory of agency and BDI that focuses on the beliefs, desires and intentions of the main actors within the phenomenon studied.

Figure 9 shows the positive relationship between the three dependent variables on the study and organizational performance. By carrying out improvements in the development of new products, innovation management and the decision-making process, the organization is expected to increase its organizational results and get success reflected according to planning developed by managers in their periodic planning.

This document provides empirical evidence on the development of new products, the management of innovation and the process of decision making as phases or variables of the process that improves organizational performance in companies in the information technology sector. This study analyzes these aspects from a systemic

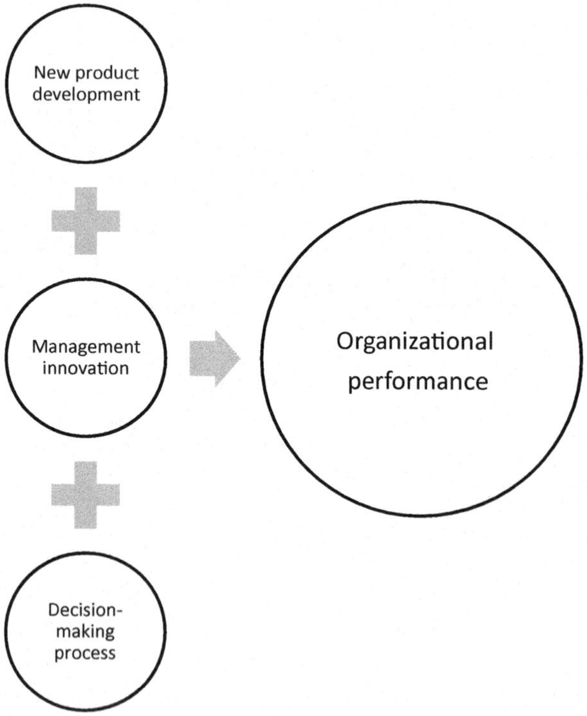

Fig. 9 Model proposed to increase organizational performance

approach that manages to make a relationship between agents in their different levels that are part of the companies studied in the technological sector.

It combines research methods by using deductive and inductive logic for the verification of the research model proposed and illustrated in Fig. 9. The results of the qualitative research provide a reliable basis for the conceptualization and operationalization of the model as well as offer useful information for the construction of the measurement instrument that allows the statistical analysis of quantitative research. The latter confirms its results through the implementation of computer simulation systems to increase the validity of the results obtained. Triangulation of methods provides to this investigation conclusions with greater methodological support.

The process of organizational performance is analyzed from a holistic approach by combining traditional research approaches with computational methodologies. The decision-making process is part of the management of the company, this item is understood to depend on the strategies promoted in the upper management to improve the approach that is implemented to develop actions supported by data and verifiable methods towards the interior of the company. In the same way, the management of innovation is an activity that improves the development and design of new products, due to this it is important that the decision-making process is a fundamental part of everything that the organization does in relation to an end strategic that improves the organizational performance.

Finally, the proposal to carry out a study with a mixed approach, including quantitative and qualitative methods and also the use of computational technologies for the simulation of the phenomenon, confirms the need to relate social phenomena with computational methods. In this case, the latter improve the validity perspective of the decision-making process in the development of new products.

References

1. Wang, Y., Ruhe, G.: The cognitive process of decision making. Int. J. Cogn. Inf. Nat. Intell. **1**, 73–85 (2007). https://doi.org/10.4018/jcini.2007040105
2. Barney, J.: The resource-based view of the firm: ten years after 1991. J. Manag. **27**, 625–641 (2001). https://doi.org/10.1016/s0149-2063(01)00114-3
3. Peteraf, M.: The cornerstones of competitive advantage: a resource-based view. Strateg. Manag. J. **14**, 179–191 (1993). https://doi.org/10.1002/smj.4250140303
4. Abubakar, A., Elrehail, H., Alatailat, M., Elçi, A.: Knowledge management, decision-making style and organizational performance. J. Innov. Knowl. (2017). https://doi.org/10.1016/j.jik.2017.07.003
5. Lepora, N.: Decision making. Oxford Scholarship Online (2018). https://doi.org/10.1093/oso/9780199674923.003.0028
6. Baregheh, A., Rowley, J., Sambrook, S.: Towards a multidisciplinary definition of innovation. Manag. Decis. **47**, 1323–1339 (2009). https://doi.org/10.1108/00251740910984578
7. MacGregor, S.: Strategic management of technological innovation. J. Prod. Innov. Manag. **23**, 102–104 (2006). https://doi.org/10.1111/j.1540-5885.2005.00184_4.x

8. Miller, D., Friesen, P.: Innovation in conservative and entrepreneurial firms: two models of strategic momentum. Strateg. Manag. J. **3**, 1–25 (1982). https://doi.org/10.1002/smj.4250030102
9. Thi, N., Thuong, H., Li, Z., et al.: Advances in intelligent systems and computing. In: Proceedings of the Tenth International Conference on Management Science and Engineering Management (2017). https://doi.org/10.1007/978-981-10-1837-4
10. van Riel, A., Lemmink, J., Ouwersloot, H.: High-Technology Service Innovation Success: a decision-making perspective. J. Prod. Innov. Manag. **21**, 348–359 (2004). https://doi.org/10.1111/j.0737-6782.2004.00087.x
11. Huarng, K., Mas-Tur, A., Calabuig Moreno, F.: Innovation, knowledge, judgment, and decision-making as virtuous cycles. J. Bus. Res. **88**, 278–281 (2018). https://doi.org/10.1016/j.jbusres.2018.02.031
12. Christiansen, J., Varnes, C.: Making decisions on innovation: meetings or networks? Creat. Innov. Manag. **16**, 282–298 (2007). https://doi.org/10.1111/j.1467-8691.2007.00441.x
13. Dane, E., Pratt, M.: Exploring intuition and its role in managerial decision making. Acad. Manag. Rev. **32**, 33–54 (2007). https://doi.org/10.5465/amr.2007.23463682
14. Yolles, M.: Organisational intelligence. J. Workplace Learn. **17**, 99–114 (2005). https://doi.org/10.1108/13665620510574496
15. Castanon-Puga, M., Flores-Parra, J., Gaxiola-Pacheco, C., et al.: Extending NetLogo with interval type-2 fuzzy inference systems. In: 2014 Second World Conference on Complex Systems (WCCS) (2014). https://doi.org/10.1109/icocs.2014.7060897
16. Ahumada-Tello, E., Castanon-Puga, M.: Modelling complex systems with distributed agency and fuzzy inference systems. Knowledge-based curricula in higher education. Procedia Comput. Sci. **80**, 2317–2321 (2016). https://doi.org/10.1016/j.procs.2016.05.429
17. Márquez Lobato, B.: Metodología Multinivel utilizando Sistemas Inteligentes, 1st ed. 200 (2013)
18. Castro, J., Castillo, O., Melin, P., Rodríguez-Díaz, A.: A hybrid learning algorithm for a class of interval type-2 fuzzy neural networks. Inf. Sci. **179**, 2175–2193 (2009). https://doi.org/10.1016/j.ins.2008.10.016
19. Bock, A.: The concepts of decision making: an analysis of classical approaches and avenues for the field of enterprise modeling. In: Lecture Notes in Business Information Processing, vol. 235, pp. 306–321 (2015). https://doi.org/10.1007/978-3-319-25897-3_20
20. Rantala, J., Koivisto, H.: Optimised subtractive clustering for neuro-fuzzy models. In: 3rd WSES International Conference on Fuzzy Sets and Fuzzy Systems (2002)

From Simulation to Implementation: Practical Advice for Policy Makers Who Want to Use Computer Modeling as an Analysis and Communication Tool

Javier Sandoval Félix and Manuel Castañón-Puga

Abstract The public policy design process is an exercise that has become increasingly complex, involved in great uncertainty that can fall into improvisation and ignoring long-term effects. Thus, the possibility of decision makers to experiment and discover through computer models the possible results of public policies before they are implemented would be a helpful tool to have, generating useful anticipatory knowledge and non-reactive strategies. Nevertheless, institutional planning exercises that make use of these models are practically absent despite the existing tools to elaborate them, given that until now it has been almost an activity restricted to the scientific community due in part by the lack of introductory literature focused to the non-specialized public servant. In light of this situation, this chapter shows basic computational modeling concepts and advice aimed at policymakers presented from an urban planning perspective. It starts by showing a fundamental weakness of current policy design: ignoring the relationships and interactions between the components of an urban context. Next, the complexity approach is presented as a way to address this absence, along with the methods and tools that it uses, and the application of this approach on public policies and cities through computational modeling. Finally, a series of considerations and recommendations are presented regarding computational model-building. Based on recent model communication protocols of the scientific community, the chapter concludes with a model presentation guide aimed at a non-specialized audience, as it may be a necessary step as a part of the policy implementation process.

J. S. Félix (✉)
Instituto Municipal de Investigación y Planeación de Ensenada, Calle Tercera No. 1323-13 Plaza Elva Zona Centro, Ensenada, BC, Mexico
e-mail: construsol@hotmail.com

M. Castañón-Puga
Facultad de Ciencias Químicas e Ingeniería, Campus Tijuana, UABC, Calzada Universidad #14418, Parque Industrial Internacional Tijuana, 22390 Tijuana, Mexico
e-mail: puga@uabc.edu.mx

M. A. Sanchez et al. (eds.), *Applied Decision-Making*, Studies in Systems, Decision and Control 209, https://doi.org/10.1007/978-3-030-17985-4_6

1 Introduction

The public policy development process is an exercise that is increasingly becoming more transparent as the desire of citizens to participate in decision-making grows. The emergence of planning bodies with open citizen participation has been a response to this concern that has given varied results, but that undoubtedly point to the fact that decision-making in urban matters has become both more enriched but also complex. The diversity and contradiction of citizen proposals, urgent deadlines, the shortage of municipal human and economic resources and the presence of political interests make decision making and design of public policies an exercise involved in great uncertainty that can easily fall into improvisation and short-termism ignoring the long-term effects that can happen in the city. It is therefore important to have elements of judgment as informed as possible about the possible impacts when making a specific decision. Telling decision makers and those who influence them about possible future outcomes of their choices would be an approach that could generate anticipatory knowledge and non-reactive strategies, as usually happens. Moreover, a "public policy mock-up" would be a useful tool to have on which stakeholders could experiment and discover by themselves the possible consequences. Currently, institutional planning exercises in the presence of anticipatory knowledge produced by these mock-ups, or models, are practically absent from local government planning processes despite the existing tools to elaborate them.

The topic of public policy design and testing will be presented in this chapter, taking urban development as an example. Weaknesses of the current form in which they are designed are given, how this process is being improved through the complexity approach, and how computational modeling, as one of its tools, can be applied for its improvement. Weaknesses focus on explaining that current design processes fail to consider the interrelationships and interactions between components that public policy wants to influence, how the complexity approach addresses ways of studying these interrelationships and how computer modeling facilitates their processing and study through Agent-Based Models. The chapter concludes with advice focused on planners who want to start modeling public policies and with a proposed modeling presentation guide for the non-specialized audience based on recent model communication protocols.

1.1 The Challenges of Public Policy Design

A significant challenge in the context of the development of cities is how to design better public policies in an environment of urban uncertainty and unawareness, thus persisting a separation between proposals and realities [1]. In an institutional context where urban development programs with a normative, static and top-down vision have not been sufficient to reach its own urban goals, there is a need to improve the processes on how these programs are created and on how to generate knowledge

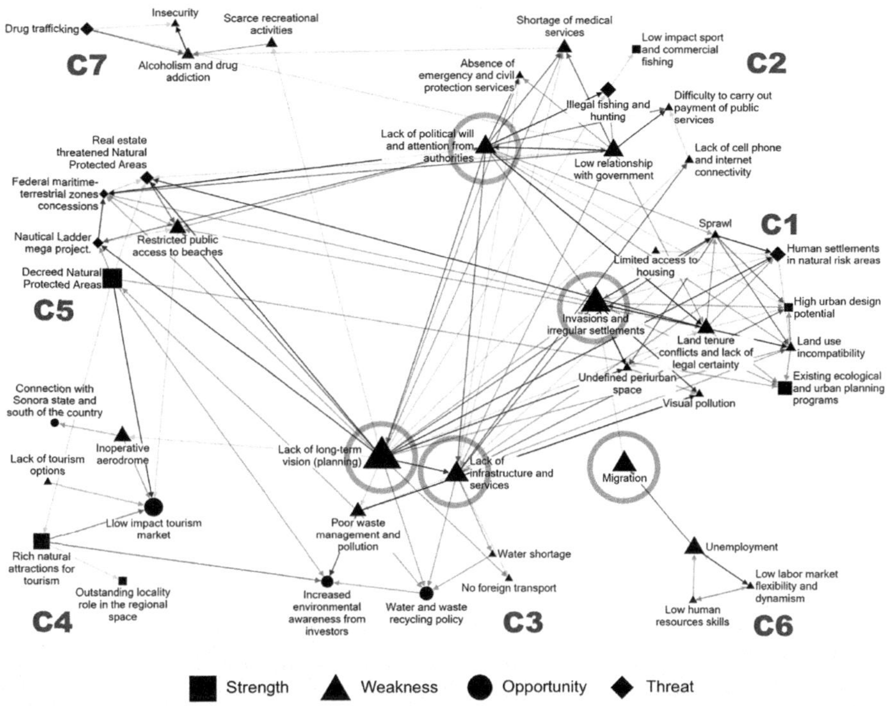

Fig. 1 Example of a network theory-based diagnosis: clustering algorithms and betweenness centrality measures to identify critical issues of a heterogeneous urban-rural system in Bahía de Los Ángeles, Ensenada, Mexico [6]

that sheds new light on how cities function. It implies overcoming today's current planning perspective that defines the city as predictable, controllable and plausible [2], as an engineering problem and consequently attending it like a machine [3]. Under this conception, current planning processes have focused mainly on descriptive, static and form aspects, leaving out a critical element to understand cities: the relationships and interaction between its components (Fig. 1) [4, 5].

This urban knowledge must start from the recognition that cities are the result of not only institutional planning efforts, but also of a large number of independent actions, organized and interacting with each other and their environment that escape governmental control efforts. Social movements, land invasions, stakeholder influence, transportation, and sectoral interests are a sample of agents that elude governments urban planning efforts but influence a city's state. Their intense interaction results in unexpected and novel outcomes such as spontaneous commercial corridors, road congestion, or the dawning of an industrial vocation. Apart from this internal dynamic that shapes cities, there are also external factors that are uncontrollable. By depending on foreign economies or joining regional service networks, a city must be conceived as an open system. Sudden services interruption due to natural phenomena or international economic oscillations can be a turning point that determines a new direction for cities.

Due to this context of interacting factors at different scales that can lead to unexpected results, is that it is essential to focus on public policies under an approach that contains the tools and methodologies that can address this complex nature. Beyond the common sense that can advise the correctness of its application, there will be public policies that are implemented for the first time, and therefore the ramifications of its impacts are unknown. Institutions that have studied the effect of public policies such as the Organization for Economic Cooperation and Development assert that small changes in public policies can cause great effects, big changes can have surprisingly small effects and effects can come from unanticipated causes [7]. Even if they were successful at the time of their application, it could be that in the long term public policies could be counterproductive [8]. It is also not uncommon to propose reacting strategies to these possible adverse effects since, apart from the fact that they have not been applied, there is not enough knowledge about the cities where they will be applied.

This lack of understanding can lead to dangerous situations because if those responsible for formulating public policies are attracted to improvising actions without knowing the nature of the problem, there is a good chance that they may worsen it [9]. Before designing an urban public policy, the underlying mechanisms of the city must be understood to identify which is the best way to induce it towards the desired path, even if an action seems counter-intuitive or implies not reaching an optimal solution [8]. Recent works carried out by the Municipal Institute of Research and Planning of Ensenada, Mexico, (IMIP) confirm the above. In the study titled Geo-statistical Study of Crime Phenomenon [10], relationships between certain types of crime were identified with the application of government actions that were initially conceived as beneficial, but results were obtained beyond what could be expected. Such is the case of the construction of sports facilities that reduce the deficit public spaces in marginalized areas, but at the same time raise the opportunity for car theft by the parking surface that always accompanies this type of projects. Or when new employment opportunities gave rise to domestic robbery since beneficiaries now had to leave their home during working hours. Another study was the Partial Urban Development Program of Maneadero-Punta Banda [11]. When designing its land use regulations, it was stipulated to respect an agricultural valley, prohibiting any type of urbanization. The benefits that this would bring to the economic activity of agriculture were never doubted, but when simulating this situation in a computational model [12], the results showed that if this public policy was implemented, there would be a possibility that urbanization could happen in risk areas outside the agricultural valley (Fig. 2). The model clearly showed that by acting on one space, unintentionally could affect another originally not intended to. Designing a public policy on the valley implied considering its effects beyond its borders. In light of this findings, it would be helpful to carry out urban research from the perspective of interacting elements that make up the city and thus provide informed support for the design of public policies on urban development.

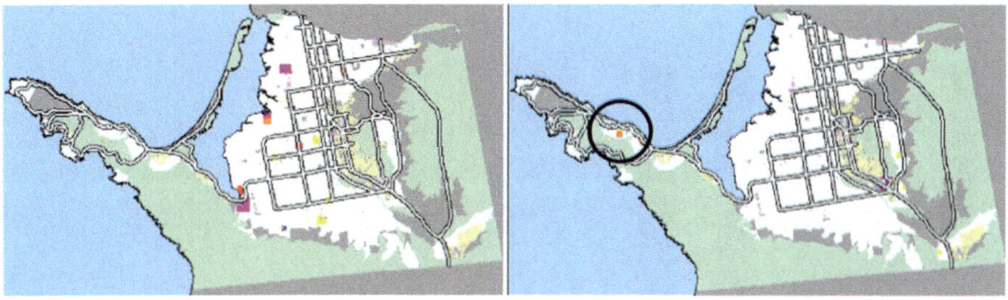

Fig. 2 Left: scenario without applying policies where urbanization of the agricultural valley is observed. Right: scenario of application of public policy to avoid the urbanization of the agricultural valley. New urbanization that is generated outside of it is enclosed in a circle [12]

2 Complexity and Emergence

Affecting areas that were not contemplated in a public policy is evidence of a lack of knowledge about relationships between the policy's target, its environment, and other areas that can be seen as disconnected at first glance. This web of relations is encompassed within the term of complexity, a word that comes from "complectere" of Latin origin whose root "plectere" means intertwined (Fig. 3). Its genesis goes back to systems theory, or the interdisciplinary study of cohesive conglomerates conformed by interrelated and interdependent parts, which had an outstanding leap in the middle of the last century. It happened when Warren Weaver, director of the Division of Natural Sciences of the Rockefeller Foundation, raised the question about what science really was in a 1948 article, and with his answer ended up writing one of the most influential texts in the field of complexity [13]. Its value resided in writing clearly and for the first time the fundamental conceptual aspect of complexity: the interaction of a large number of components. Weaver explained how science for the last three hundred years had given solutions to problems of physical nature based on the management of the number of variables, from two or three variables three centuries ago to hundreds or thousands at the beginning of the 1900s. But explanations could not be achieved when problems were related to life sciences if a particular characteristic was not included: organization. Weaver coined the term Problems of Simplicity to those that were solved based on two or perhaps a few more variables, and Problems of Disorganized Complexity to problems where hundreds or thousands of variables were involved. Regardless of the number, the components of the two previous concepts behaved individually erratic, but precisely because of their independent behavior they could be studied with probabilistic methods and statistical mechanics, and thus know the likely behavior of the system under study. For systems with interrelated components in an organic whole, Weaver coined a third term: Problems of Organized Complexity (Fig. 4). Due to the intense interconnectivity and interaction of the components of this type of systems, the author came to visualize the need to use multidisciplinary working groups and the processing power of computers for their study and understanding, something visionary at that time.

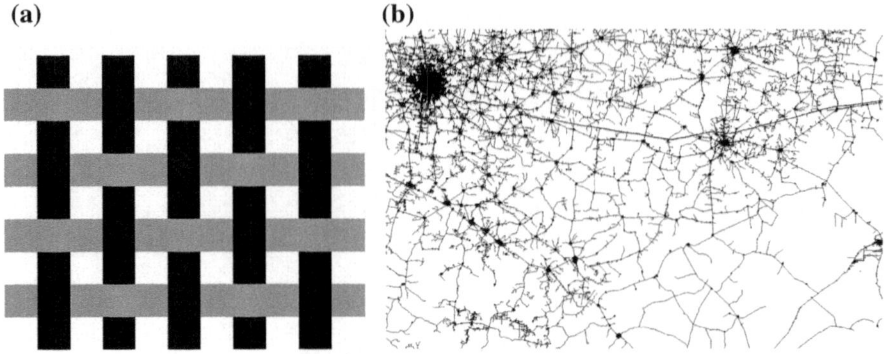

Fig. 3 **a** The word complexity comes from the root "plectere," intertwined: a fabric of inseparable elements that constitute a whole. **b** Road network of the state of Yucatan, Mexico

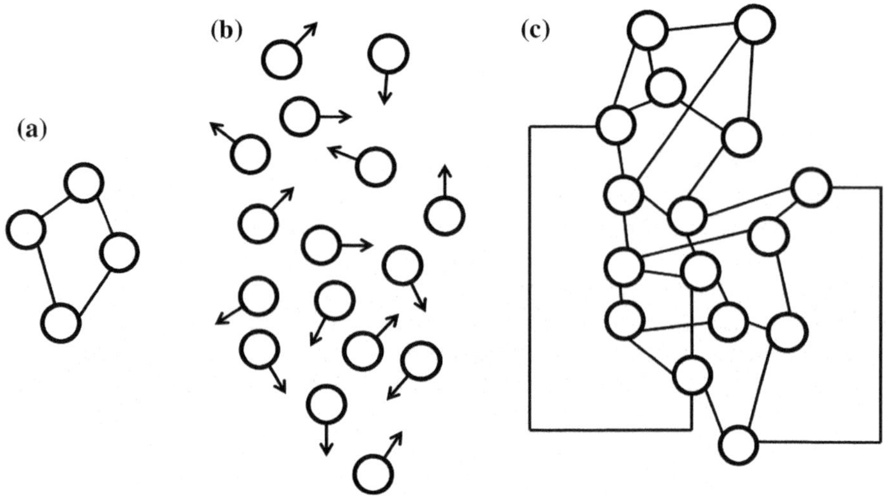

Fig. 4 **a** Simplicity, **b** disorganized complexity, **c** organized complexity

Since then, given the central role interconnectivity plays in its definition, the complexity approach has been used in fields that manifests a great interrelation between its constitutive parts, like financial markets [14], epidemics [15], social networks [16], vehicular traffic [17], crowds [18], public policies [19], cities [20], and any another field that can be abstracted as a system composed of a large number of interconnected and interacting components. For this reason, the complexity approach, or complex systems, has not been exclusive of a particular science but has functioned as an inter/transdisciplinary exploration at different scales and environments [8], regardless of traditional borders between the human and natural sciences [21]. Because of this broad application there is no consensus on its definition, but a complex system can be understood as "a system in which large networks of components

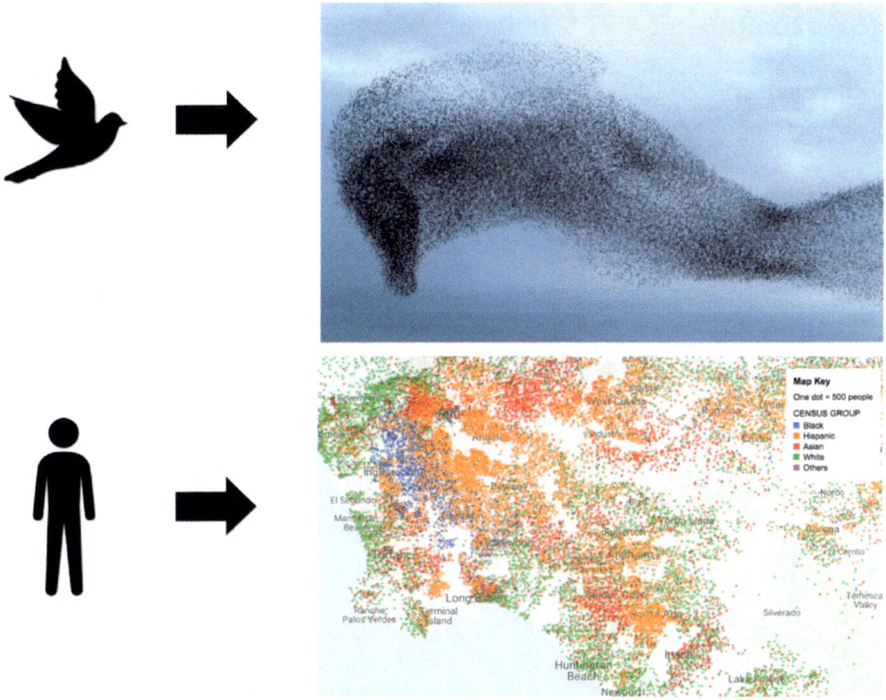

Fig. 5 Macro-behaviors that emerge from individual micro-behavior of a system components. Above, left: a bird as a basic component. Above, right: from the interaction of hundreds of birds, a flock emerges that moves in a coordinated manner as a single unit. Bottom left: a human being as a basic component. Bottom right: hundreds of individual decisions regarding residence selection, as a whole generate macro patterns of racial segregation in the city of Los Angeles. *Image source* https://conceptodefinicion.de/wp-content/uploads/2017/12/Parvada.jpg and https://www.nytimes.com/interactive/2015/07/08/us/census-race-map.html

with no central control and simple rules of operation give rise to complex collective behavior, sophisticated information processing, and adaptation via learning or evolution" [22]. This collective behavior is called emergent process and refers to the rise of new and coherent structures, properties and patterns during the process of self-organization [23]. This unpredictable behavior is the result of interactions between a system's components (Fig. 5). When the behavior of a system is described with information different from that used to describe its components [24], it is considered an emerging process, where the whole is not only more than the sum of its parts but also different: more is different [25]. It can be said that the complexity approach is characterized by being anti-reductionist (if components of a system are eliminated, its consequences cannot be known with certainty), holistic, multidisciplinary (due to the heterogeneity that may exist between the components) and non-linear (the sum of the components is not equal to the system of which they are part). Its research topics, methods, and tools are shown in Table 1.

Table 1 Features, research topics, methods and tools of complexity

Features	Research topics	Methods	Tools
Anti-reductionist	Self-organization	Cellular automata	Computer equipment
Holistic	Adaptation	Chaos	Modeling and computer simulation
Multidisciplinary	Emergent processes	Graphs	Programming languages
Non-linear	Autopoiesis	Percolation	Computer software
Multiplicity of states	System dynamics	Information theory	
	Evolution	Neural networks	
	Networks	Fuzzy logic	
	Phase transition	Genetic algorithms	
	Performance according to scale	Fractal geometry	
	Learning	Artificial life	
	Information processing	Network analysis	
	Path dependence	Multi-agent modeling	
	Far from equilibrium dynamics	Data mining	
	Patterns	Artificial intelligence	
	Segregation	Games theory	
		Numerical simulation	
		Power laws	
		Machine learning	
		Monte Carlo simulation	
		Markov chains	
		Differential equations	

70 years after the publication of his influential paper, Warren's vision has not deviated from the three current trends of complexity. First, interdisciplinarity and analogy. Second, the techniques and models that emerged from operations research and computational simulation. Third, and perhaps one of the main differences between a system and a complex system, the notion that the optimization of a system through its design and control is no longer an adequate metaphor for solving human problems [26]. This last aspect caused urban disciplines to search for new approaches to city problems through complex systems thinking as central and mechanistic systems of planning declined.

2.1 Complexity and Public Policies

As already mentioned, the complexity approach has been applied in several fields of natural and human sciences including cities and the public policies that try to control and induce them towards desired states. As policies and complex systems share the interest for the agents involved and their interactions [27], it is thus feasible to apply complexity's research topics and methodologies. They offer new ways of designing public policies, focusing on dynamic connections and evolution, not only on the static design of institutions, laws, regulations and other traditional instruments [16]. Beyond causes and their control, they highlight the importance of influence and probability. The complexity approach overcomes the traditional way of designing public policies by taking into account that these have effects in other domains that interact with each other [8], that interactions with apparently unrelated domains are unknown [28] and that they will operate in a highly complex space beyond what can be controlled or observed in a deterministic way [29]. Therefore, the complexity approach identifies new relationships and interactions in urban systems in a context of uncertainty that allows more sustainable and successful solutions [30]. It draws knowledge from cities by revealing underlying mechanisms that lead to observed behavior [7], as well as identifying emergent behaviors and anticipating policy impacts before their application in real life [31]. Thus, the goal of public policies would be met: to alter or maintain the behavior of a large group of individuals or organizations to achieve a desired social outcome. The goal of public policy analysis, which is the systematic comparison and evaluation of alternatives available to public actors to solve social problems, would also be met [8]. New insights for urban planning could be achieved to stimulate or inhibit the emergence of new urban qualities [32], improve a system's resilience, diminish its vulnerabilities, inhibit dangerous tipping points and identify key actors that can promote changes in the system [8]. Thus, forecasting, explaining and controlling would have important conceptual implications for those who design and implement public policies [7]. This union between public policies and complexity (Table 2) is characterized by containing multiple actors and levels of government, and processing policies in different ways: from intense disputes with many actors, issues routinely dealt outside the public spotlight, to decision-making based on limited information, high levels of uncertainty and ambiguity [33]. Even so, and despite the potential usefulness of the complex urban systems approach and the increasing accessibility to high computational processing power, there are few examples of application since there is not yet a coherent and unified set of concepts and tools for planners and public policies designers [19].

2.2 Complexity and Cities

When Warren Weaver wrote his influential article on organized complexity, he already glimpsed the application of quantitative and mathematical experimental methods to the sciences of biology, medicine and even the social sciences as a way

Table 2 Public policies research topics and corresponding complexity

Public policy research topics	Complexity methods	References
Social unrest, wars	Tipping points Spatial patterns	[34, 35]
Food crisis	Far from equilibrium dynamics	[34, 36]
Market analysis	Networks Cascading failures	[14, 37]
Interdependencies and economic risks	Networks Cascading failures	[37]
Environment and climate change	Feedback networks	[34, 38–41]
Social groups	Networks Self-organization Segregation	[39, 42, 43]
Cities	Spatial patterns Emergent processes Networks	[19, 20, 30]
Transportation planning	Networks Game theory	[44, 45]
Legislative activity	Pattern emergence	[46]
Education	Multi-Agency Networks	[47]
Epidemiology	Networks	[34, 48]
Institutions, planning	Path dependency	[6, 49–51]

to study their complex nature. And although this also happened for the subject of cities, at first it was only applied by mathematicians and physicists who were more interested in carrying out tests using the cities as case studies, than to advance in the knowledge of human settlements. Urban models were related to complexity topics such as dynamic physical systems, phase transition, tipping points, far from equilibrium states, scale laws, networks, exponential growth, logistic growth, non-linear behavior, chaos and bifurcation [20]. Particularly, there were models that showed how new activities could appear spontaneously in unexpected places, as it happens in complex systems regarding uncertainty and emerging processes. This was in direct contrast to the rational comprehensive urban planning approach that assumed that cities were essentially predictable and therefore controllable and plannable [52]. By the end of the last century this approach would soon give way to a world that was increasingly conceived from the bottom up, which is a feature of complexity. By applying physics concepts to urban centers, the links between complexity theories and cities were reinforced since the 1960s both in the USA at a time of major infrastructure interventions, and in Latin America in an intense urban growth period [8]. It arose in the US partly as a reaction to the vision of solving urban aspects from purely infrastructure terms, and it was the influential activist and author Jane Jacobs who reacted against this mechanistic way of conceiving cities that annihilated their heterogeneity and diversity. In her book The Death and Life of Great American Cities [53], Jacobs uses Warren's concepts directly, arguing that urban problems could not be approached either as simple or as disorganized complexity, but rather as an organized complexity, proper of the life sciences. This implied that cities were not to be seen as machines, but with characteristics of living systems from which the shape of the city emerges from the bottom up [54]. Her book brought forward the perspective of complexity of cities, generating interest and participation not only of physicists and mathematicians but of sociologists, architects, and urban planners. It contributed to the formation of a new domain of study that has been named Complexity Theories of Cities [55], which explores a variety of urban phenomena that were previously considered independent and studied under different theoretical bases [2], but which are now considered interconnected in the intricacy of urban complexity [32]. Complexity Theories of Cities exceeds the limits of traditional urban planning and unites aspects of cities and complex systems, where population and economic growth, increased levels of urbanization, dependence on infrastructure and the role of technology in society are considered elements that drive urban complexity [56] achieving a solid theoretical base for a variety of urban phenomena and properties, as well as providing new understanding of cities with basic features of complexity [2]. Given that cities are composed of interacting and interrelated elements as a complex system, this approach has been considered crucial for understanding the characteristics and dynamics of urban systems regarding natural resource management [38], detection of socioeconomic patterns [57], urban landscapes [58], traffic and transport [59, 60], connectivity [20], population [18], epidemics [61], land use changes [62] and social networks [16], among other topics (Table 3).

Table 3 Complex urban systems research topics and methods

Complex urban systems research topics	Methods	References
Land use spatial patterns	Cellular automata agent-based models	[63–71]
Spatial segregation of social groups	Cellular automata	[72, 73]
Economic performance according to scale	Power laws	[4]
Communication and services networks	Network analysis	[74]
Social networks	Network analysis	[75–78]
Urban emergent processes	Cellular automata Multi-agent modeling Neural networks	[79–82]

(continued)

Table 3 (continued)

Complex urban systems research topics	Methods	References
Positive feedback based development	Network analysis Multi-agent modeling	[83]
Social self-organization	Network analysis Multi-agent modeling	[2, 15]
Phase transition towards new system states	Montecarlo simulation	[4, 84, 85]
Transportation, traffic and congestion	Network analysis Graph theory Game theory	[44, 45, 60, 86, 87]
Pedestrian mobility	Network analysis Speed and density flow equations Graph theory Percolation theory	[15, 18, 88]
Crowds, pedestrian emergency evacuation	Network analysis Percolation theory Graph theory	[88]
Residential mobility	Cellular automata Multi-agent modeling	[27, 89, 90]
Commerce spatial location	Cellular automata Multi-agent modeling	[91]
Health and infectious vectors	Network analysis	[84]
Land market and real estate interactions	Cellular automata Multi-agent modeling	[14, 18]
Slums	Cellular automata Multi-agent modeling	[81, 92–94]
Sprawl	Cellular automata Multi-agent modeling	[95–97]
City size rank	Power laws	[4]
Urban planning processes	Multi-agent modeling Neural network Genetic algorithm Network analysis	[57, 59, 98–105]
Urban disasters contingencies	Multi-agent modeling Network analysis	[7, 56, 106]
Spatial distribution of economic aspects	Cellular automata Network analysis Multi-agent modeling	[14, 42]

What initially began as case studies for other disciplines driven by physicists and mathematicians, over time has fertilized the field of urbanism, updating it to the conception of the city with behavior and evolutionary organization where it changes and adapts [107], its classification is carried out according to power laws [108], a genetic code is identified based on functional and spatial structures [109] and is conceived as a hyper-network or "system of systems of systems" [110], among other aspects. The field of complexity can shed new light on innovative methods for the study of human settlements [16] since the understanding of their functional relationships is one of the critical factors for the design of effective public policies of urban development [57]. Despite these advances, the presence of professionals in the field of urban planning and planning is still nascent, so the contribution of complexity to the study of cities is just beginning.

3 Modeling

There is a seemingly insurmountable obstacle regarding urban knowledge generation. It is the scale, cost and time that urban experiments would require if they were carried out in real time and size. The cornerstone of the scientific method is experimentation and its independent reproduction under controlled conditions. From a cities perspective, an urban laboratory would cover an area of thousands of hectares on which a human settlement would be built to test a theory. Although experiments have been carried out on this scale [111], they can only be financed by extraordinarily solvent economies. Moreover, although they are known as experimental cities, their focus has been on innovation in the sense that something is done for the first time, not for scientific experimentation nor to having to think about the consequences, or resistances, of the population. Considering only the possibilities of continuing with an experiment after what could be considered a failure, these are very different between a real city and what could be done in a laboratory. Fortunately, apart from conducting in situ tests, there is the most accessible option to perform them in silico, through a computational model.

In the broadest sense of the word, a model is a simplified representation of a structure or system [24]. Under this concept, there are several ways of representing cities, but all can be included in physical and conceptual forms of representation [112]. In the former, a city is represented through iconic physical or analogs (plans, maps) models, and in the latter through statistical, mathematical, theoretical or computational models. Given the demands of information processing of complex systems, their representation and work are given through computer equipment, which corresponds to the conceptual representation or scientific modeling [113] through computer modeling.

3.1 Computational Modeling

Given a large number of variables, states, interactions, relationships, and rules to be considered in a complex system, it is asserted that without the existence of computers the field of Complexity would not have emerged [114, 115]. With the formalization of an urban system through a computer model, it is possible to do a series of exercises to understand it and predict its possible future states that for urban decision makers can be of great importance. Since the 1960s, with an initial focus on urban transport and land use issues, the processing power of computers has been used to implement complexity research topics through methods such as cellular automata (Fig. 6), [49, 95], chaos theory, neural networks, fuzzy logic, genetic algorithms, fractal geometry, artificial life, network analysis and multiagent models to perform models and simulation exercises that contribute to the understanding of urban centers or that forecast their future behavior [114]. This methods application has been varied: hybrid logistic regression models and Markov chains to simulate population growth, geographic automata [97] to represent drivers of dispersion, micro-dynamic processes [96] to study dispersed growth through links of macro-scale processes, probabilistic maps of scattered land uses [116] and their modeling in Geographic Information Systems [117]. Such has been the recognition of the potential of simulation that has been classified as the third way of doing science in addition to the inductive and deductive processes [118], and has come to be called as Computational Social Sciences [119], understood as "… the integrated, interdisciplinary investigation of social systems as information-processing organizations and through the medium of advanced computational systems." [120] (Fig. 7).

The usefulness of these tools lies in the fact that "what if" questions can be asked as a way to analyze public policies [9] where they can be tested in silico with different options in a practical and economical way to know its effects, effectiveness, and costs [107]. As mentioned by Furtado et al. [115]: "working with complex systems applied to public policy means to create computational experimental environments in which the essence of the systems is present and from which one can withdraw elements of improvement of public policies in a relatively simple and cheap way." Identifying the key mechanisms of an urban system is essential since it can help determine how it should be affected to cause a particular change and this, together with avoiding unwanted surprises, is precisely the goal pursued by public policies. Without this understanding, it would be more challenging not only to know what "lever" would have to be used to obtain the desired effect but also when it would have to be stopped so as not to start having undesirable effects [8]. In an environment of limited budget and resources where large urban projects become prohibitive due to their high cost, surgical and selective intervention becomes essential. In the light of complexity, one key issue for the administration of complex systems is that the solutions are not imposed per se, but are extracted from the system itself, to guide it towards one of the available states to which it can be driven, seen from an integral point of view of its components [9].

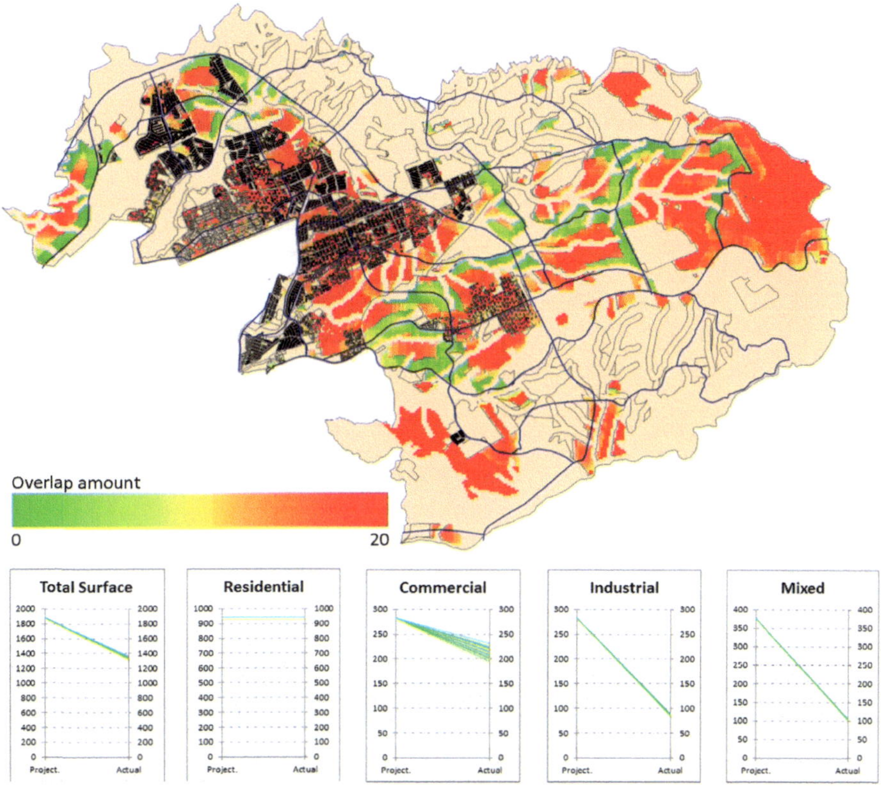

Fig. 6 Urban simulation result focused on identifying urban impacts in Ensenada, México [49]

If the complex urban systems approach through computational simulation can be used for improving public policy design, then its usefulness apart from understanding and prediction [24], also lies in planning. On the one hand, it helps in systems understanding through the characterization of their behaviors and thus identifying essential properties [8]. Even the process of model construction can be as enlightening as the use of the model itself because it reveals what we know and what we do not know about its connections and causalities [115], and with this knowledge, better public policies can be designed. In this way, one of the potentials of urban modeling would be achieved: the applicability of these tools by the social organizations in charge of governing the development of cities [119]. On the other hand, it can be used to forecast outputs that could unexpectedly result if public policies are applied. And, although modeling does not indicate with absolute certainty the effects of a policy in real life, knowing the variety of possible effects will ensure a better understanding of the whole and thus ensure more robust public policies [8]. For example, a model may suggest that for a system to reach a previously desired state, the actions needed to apply may be counterintuitive, that individual responses may

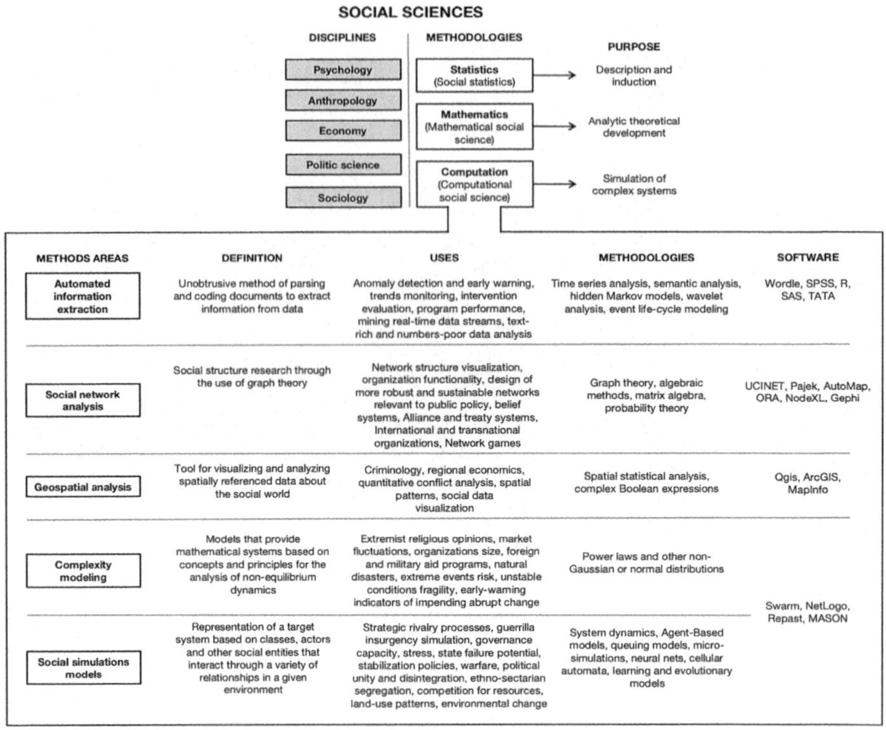

Fig. 7 Computational social sciences. *Source* Adapted from [120, 121]

differ from aggregate effects, that results are path-dependent, and that institutional details may be meaningful. These are powerful lessons for decision makers. However, beyond that, if models are not adapted to the cities interested in applying them, they will not be as useful.

3.2 Agent-Based Models

A computational modeling approach used in the field of complex systems is Agent-Based Models (Fig. 8). As its name implies, the basic unit is an agent, which is a unique and autonomous entity that interacts with the rest of the agents that exist in the model and with the environment in which they are immersed. The main characteristics of an agent are [122, 123]:

- Autonomy: There is no central control of the agents, they decide for themselves.
- Heterogeneity: Each agent can have its own attributes (states of variables and values) and rules of behavior.

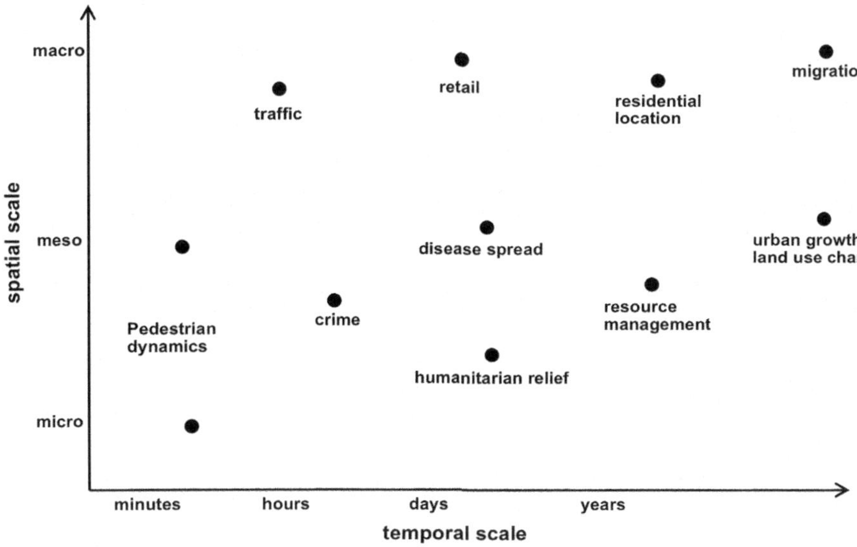

Fig. 8 Agent-based models use according to time and space scales. Adapted from [125]

- Explicit space: Agents exist within a space, which can be a physical or abstract environment (for example, a social network).
- Limited rationality: Agents do not have universal knowledge, only that which is specific to their context.
- Rules of behavior: Each agent must be able to respond proactively to changes in their environment.

An agent can be a human, an organism, a corporation or any other entity that pursues a goal [124]. Because of this flexibility of application, agent-based modeling has been used in such diverse subjects ranging from the study of pedestrian flows to urban growth and large migrations [125]. The purpose of this type of models is to create agents and rules that generate a behavior of interest. When this behavior develops in space, a family of these models is called cellular automata.

3.2.1 Cellular Automata

Complex systems, as already mentioned, have a broad application, and the selection of methods and models will depend on the fields in which they are used. There are systems that have the property of being ubiquitous, such as financial markets or social networks where people can be integrated into them regardless of their location. On the contrary, there are systems where spatial location is a determinant characteristic. Such is the case of cities, which unfold in contexts explained by geography and the notion of territory, both imminently spatial. From the beginnings of systems theory, cities quickly attracted attention because of their complex intrinsic characteristic.

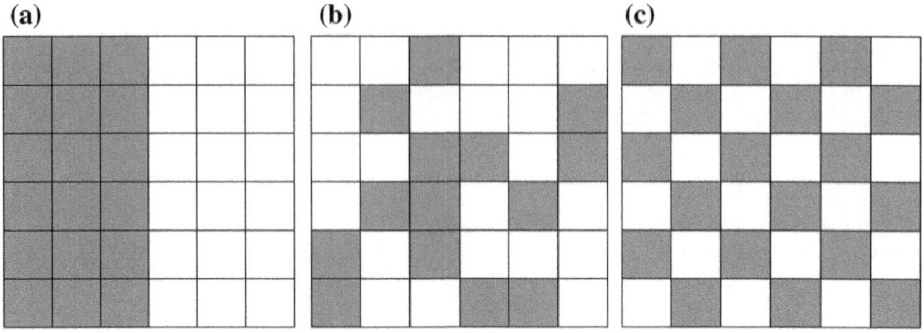

Fig. 9 Types of spatial autocorrelation: **a** positive: spatial closeness with similar attributes, **b** no autocorrelation, **c** negative: spatial closeness without similar attributes

Weaver had already written about the interrelation of variables in a general way, but it was Tobler [126] who related it to geography in such a simple way that it continues to permeate today. Accepting that in general terms everything is related to everything else, he thought that in theory, any problem could become infinitely complex, particularly in the social sciences. For example, the amount of population of a specific place would be part of the population of a particular region, which in turn would be part of the population of a country, which in turn would be part of the total population of the world. As Weaver explained that a system could be characterized as a whole through average behaviors no matter its large number of components as he exemplified with thousands of balls colliding in a billiard table; Tobler brought the example of climate to explain that it could be summarized with approximations despite being the result of variables that can be added to reach a global level. Likewise, although a social system could be considered complex due to its large number of variables, from a spatial perspective it could become more manageable if the variable of *proximity* were considered. From here came what Tobler called the First Law of Geography: "Everything is related to everything else, but near things are more related than distant things". Today this law is still cited in spatial studies and has given rise to the concept of spatial autocorrelation (Fig. 9), which is a property that measures the similarity between near and far things [127].

At the same time that Tobler unveiled his First Law, Scientific American magazine published what looked like a simple and inconsequential mathematical game of spatial nature in its column of mathematical games [128]. To be carried out on a grid sheet with chips that were on hand or even with pencil and eraser, the game was straightforward to explain and understand, but its rules were deliberately designed so that its future results could not be predicted, what was true until very recently only thanks to the use of computers. The game started by distributing chips, one inside each cell of the grid to form a pattern to suit the player likening. It was the first and only time that there was player involvement in the game since the rest was to successively apply three rules that, despite their simplicity, it continually modified the initial distribution of the chips. What happened next was one of the simplest representations of a complex system: rules and interactions that produce counterintuitive or unexpected results. Starting from the most straightforward distribution of chips, they reproduced, survived and/or died over generations spanning large areas of the grid until they reached a perpetual oscillatory or fixed state. It was not easy to know in what number of generations these states would be achieved, nor the path to get there, not even the state that the cells would be only two or three generations ahead. This game was designed by mathematician John Conway and called it Life.

After Conway's Game of Life, Tobler used the colloquial phrase "When in Rome do as the Romans do" [129] to illustrate his perspective on proximity, now with spatial mathematical models applied to geography. Nine years after introducing his First Law of Geography, he indicated that it was equivalent to saying that rules depend on the place where they are applied. This conception would be the most recent of a series of models that explained the reasons for land use change based on adjacent spatial environment (Fig. 10). He did not talk about proximity here; now he referred to a *neighborhood*, a word still used today in spatial models. In this neighborhood, *transition rules* applied in a squared space made up of *cells* that were in one of several possible *states*. Following the land use example, Tobler proposed that these states could be: Residential (R), Commercial (C), Industrial (I), Public (P), and Agricultural (A), and a neighborhood of 5 cells. There were, therefore, five states and five neighbors (north, south, east, west and central cell). Thus, a transition rule (land use change) could be (Fig. 11). Which in this case meant the change of the central cell from agricultural to commercial use. This change from one state to another could not be random but as a consequence of specific causes of each geography. The deduction of these transition rules was a topic of interest for Tobler, and it continues to be in recent studies [130–132]. He exemplified it with the following exercise: having 20 images in sequential order of the positions of pieces of a chessboard, determine their rules. He ended by asking if the same could be done for the patterns that change the earth's surface (Fig. 12).

These concepts of neighborhood, transition rules, and states managed by Tobler and Conway are characteristics of models called cellular automata. They go back to the first exercises carried out by mathematician John Von Neumann in the 40s when he studied the mechanisms of self-reproduction and life modeling, who in turn was influenced by the works of computer scientist and cryptographer Alan Turing, who worked on the creation of a computer capable of acting on its own (hence the term automaton) based on a set of parameters and knowledge of the external world.

An automaton can be defined as "a processing mechanism with characteristics that change over time based on its internal characteristics, rules, and external input" [133]. It can also be defined as "a machine that processes information, proceeding logically, inexorably performing its next action after applying data received from outside itself in light of instructions programmed within itself" [134]. Cellular automata take place in a space composed of cells, where each one processes information and acts by a series of transition rules and information received from their environment [135]. Formally, each cell A is defined by one of a set of possible states $S = S_1, S_2, \ldots, S_N$ and a transition rule set T

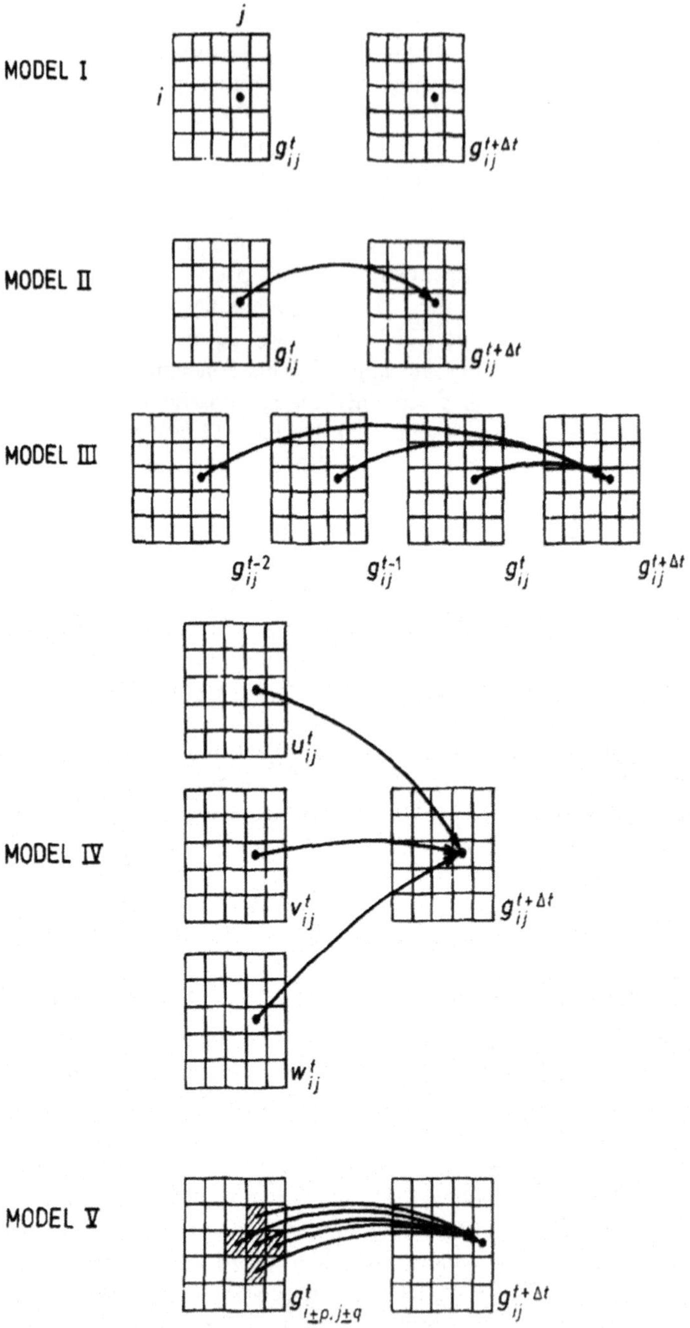

Fig. 10 Land use change models classification. (I) Independent: no relation with previous land use, (II) dependent: land use depends on the previous land use, (III) historical: land use depends on several previous land uses, (IV) multivariate: land use depends on several other variables, (V) geographical: land use depends on the land use at other neighboring locations [129]

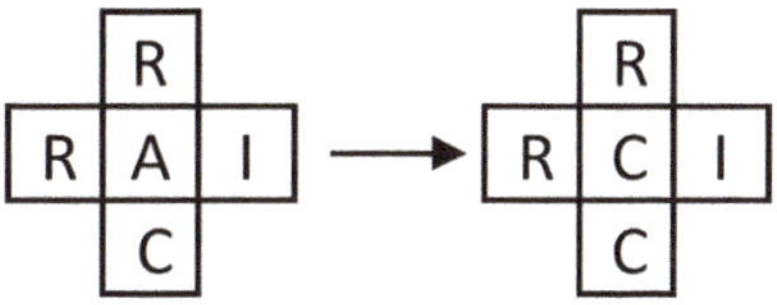

Fig. 11 Transition rule

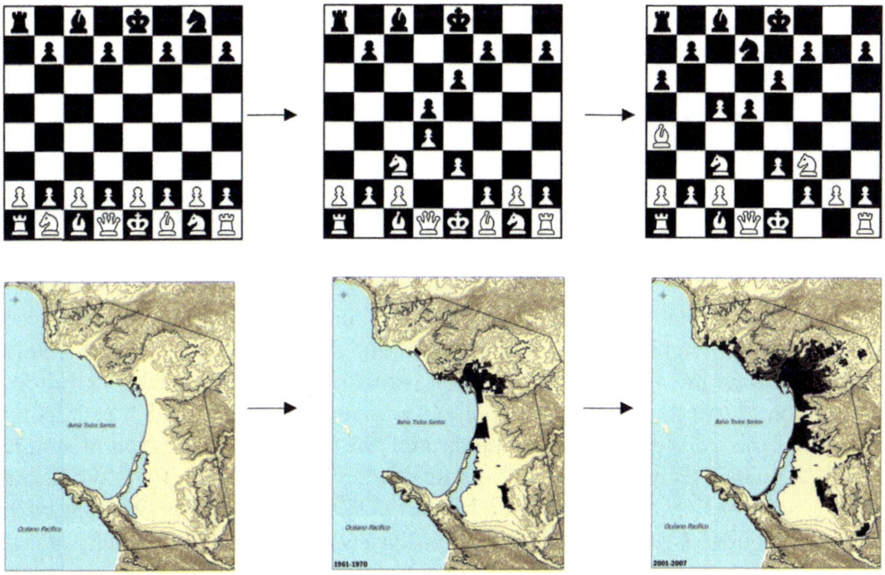

Fig. 12 Orderly sequence of a game of chess and a city's growth: can the rules of behavior be extracted by observing their evolution?

$$A \leftarrow (S, T)$$

The transition rules define the state S_{t+1} of an automaton at time $t + 1$ depending on the previous state t, $S_t(S_t, S_{t+1}, \in S)$ and the information input I_t

$$T : (S_t, I_t) \rightarrow S_{t+1}$$

information input comes from the adjacent cells states, or neighborhood

$$A \leftarrow (S, T, R)$$

where R are the neighboring cells of A, from where information I is taken to apply transition rules T for cell A. The most used neighborhoods in cellular automata are Von Neumann and Moore and may expand as in the Extended Moore neighborhood (Fig. 13).

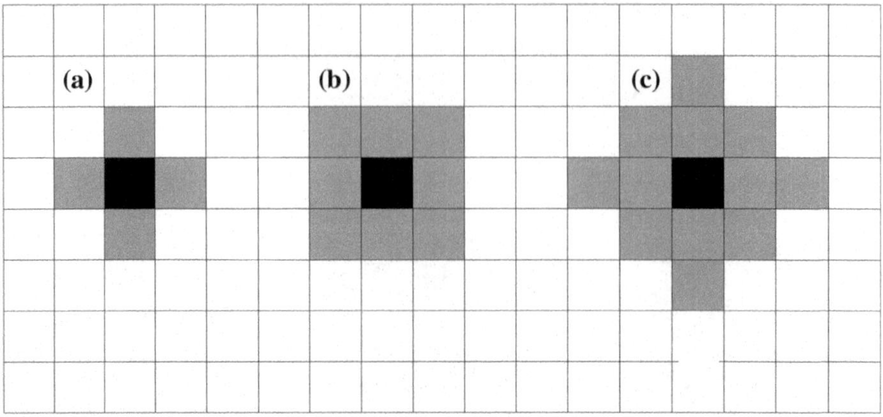

Fig. 13 Types of neighborhoods: **a** Von Neumann, **b** Moore, **c** extended Moore

Given the fact that urban characteristics like land uses are strongly influenced by their closest neighbors [52], and that cellular automata can be used to build models of urban growth patterns due to its structure, simplicity, and possibilities of evolution [136], they have been applied to urban studies [135, 137] respecting the premise that cells change their state through the repetitive application of simple rules [138] where transition rules are related to territorial factors (Fig. 14). These studies have been linked to behavior prediction [131], urban expansion [96, 139, 140], urban contraction [141], land use change [24], transportation [142], spatial patterns [143], densification processes [144] and as a tool for public policies analysis [145]. Insomuch as that there are no pre-elaborated models, and they have to be created according to the underlying complexity of the system that is being reproduced [122], a wide range of computational tools have emerged, which have been accounted for by at least 85 different platforms [146].

4 Recommendations for Beginning Modelers

The learning curve for developing Agent-Based Models can be quite pronounced for many professionals. Like all learning, it must start from the simple to the elaborate, and if it is first with an intention of free exploration without the pressure to reach a specific result, the better. In model elaboration, the process can be approached in two ways: one where we just play with the rules of the model to see what happens, called exploratory modeling, and another where we deliberately seek to reproduce a specific phenomenon, called phenomena-based modeling [123]. For someone new, it is better to start with an exploratory approach towards modeling. First, time must be dedicated to know about the general behaviors that can be reproduced in a variety of models, which can be found in model libraries that will surely be included in

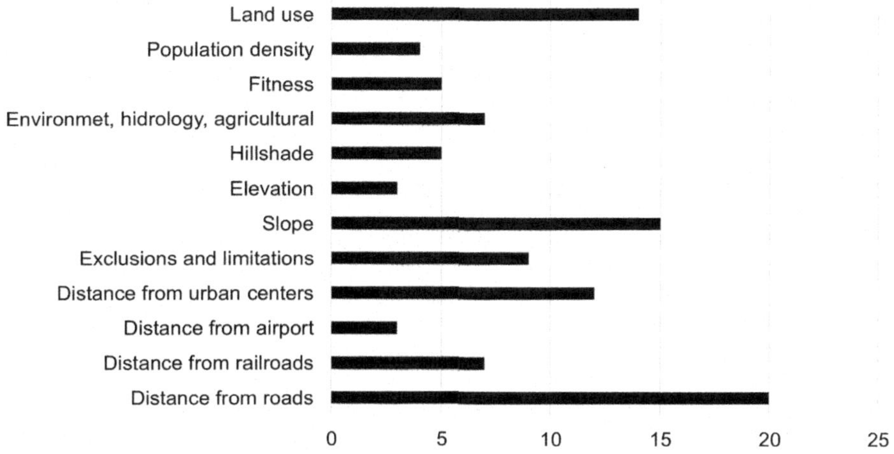

Fig. 14 Number of occurrences with which different factors appear in urban simulation studies with cellular automata. Adapted from [136]

the programming environment of choice. Once observed, models of interest must be chosen and identify their variables, input changes to these variables and observe the outputs, reason the relationships and changes between these inputs and outputs, and finally see how they are displayed. As part of this exploratory process, it is normal that, on the one hand, we get to a point where we ask ourselves questions like "What would happen if this change or that addition could be made to the model and that it currently does not have?". And on the other, that we begin to imagine the variables and behaviors of the model as entities of the real world. Although the elements of a model do not per se represent concrete aspects of reality, the latter can be abstracted at a level that can be reproduced and studied in a model. The beginning modeler will do this with more technical basis if he goes through an exploratory phase first.

This section presents a series of recommendations aimed at planning professionals who are starting to develop Agent-Based Models, based on first time experiences using cellular automata to build urban growth models as tools to aid public officials make more informed decisions about land use policies.

4.1 *The Modeling Process*

It must first be understood that a model is a simplified representation of reality. This concept demands explicitly specifying what will be the phenomenon to study along with an explanation of what are its causes are and how they relate to each other. So, before writing the first line of code, there must be a series of assumptions that explain the observed real-life behavior to reproduce on a computer model. This obliges to an acute exercise of observation and analysis of the phenomenon of interest,

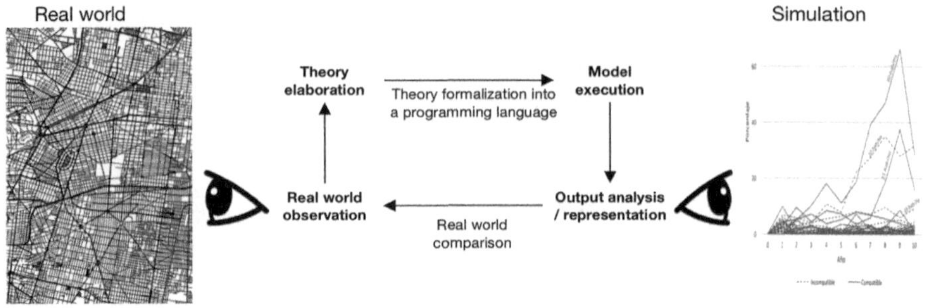

Fig. 15 Modeling process

complemented by evidence and research carried out preferably by several disciplines. Later, when a theory is built, it will be translated into a computational model, run it and verify if the behavior observed in real life was reproduced.

Possibly the first modeling attempt will be successful in reproducing the phenomenon, but it is more likely that for every success there will be numerous attempts that yield results far from intentional behavior. For this reason, modeling is a feedback process that takes the following steps: observation of the real-life phenomenon → theory elaboration that explains it → theory formalization into a programming language → model execution → output analysis and representation → real world comparison, and repeat (Fig. 15). This process is both absorbing in time and effort, but by its very nature offers a wide field of opportunities to study and experiment the phenomenon of interest. And here is one of the primary values of this methodology: the modeling process serves as an urban thinking tool, so the very process of constructing the model can become as revealing, or even more so, than its outputs. This is particularly true when making models for the first time. It is of great value to ponder about the factors that determine the dynamics of the phenomenon of interest, code the resulting assumptions, observe the outputs, realize that the intentional behavior was not obtained, revise the code accordingly, ponder again about the causes of the phenomenon and put them back to the test. At this point, it is probable that the research question has not even been addressed. In the process, other paths of research can be discovered. More than a circular process of feedback, it takes bifurcations: it gets to a point where it becomes attractive to try two or three variations and then proceed only with one or even none, which implies backtracking, move to another direction and again discover new research paths. Modeling takes a branched structure composed of a multitude of abandoned lines of work in favor of a few models.

4.2 Real World Observation

Accustomed to making descriptive and static diagnoses, for most public planning officials it is still alien to identify the interrelations of an urban system, as well as to describe how it changes over time. Unlike traditional urban analysis that maps the current state of things in a fixed cut in time, making a computational model requires identifying the components that cause a phenomenon of interest, and how they interact with each other and the environment. In this way, the first step is to describe the behavior of the phenomenon that we are interested in reproducing, as well as a range of possible causes. It is also important to identify the behaviors that a model must not output or are not of interest, but that can be unintentionally reproduced and think how can it be accidentally coded. This is important because it is a tactic to anticipate possible undesired results and to code accordingly. For example, when the first versions of urban growth models where built at the IMIP, they outputted the correct amount of projected surface, but its spatial patterns where far from those observed in the real world as there was only a random location function working in the location of new urban growth, without accounting for other factors that are involved in site selection that happens in real life. Later modeling efforts took into account evidently unrealistic pattern results, even those that no model previously outputted but none the less where a possibility by incorrect coding, so this accidental coding was identified and thus avoided.

4.3 Theory Elaboration

In 1999 the epidemiologist and social models specialist Joshua M. Epstein wrote an article in favor of the use of agent-based models in the social sciences, coining the phrase "If you didn't grow it, you didn't explain it" [147], as a way to explain that the possibility existed of reproducing and explaining social phenomena in-silico. Now, among colleagues, they would not ask "can you explain it?" but "can you grow it?" Although it is not correct to assert that reproducing a phenomenon means that the appropriate assumptions are being used, even so, the phrase raised a flag that warned of the need to have an explanation theory before modeling.

For example, in a traditional urban planning context, an attitude that is detrimental to the development of urban models is the triviality that certain aspects of the city are considered, but that the computational modeling process demands to observe and explain. For planning professionals, it is not a novelty that cities start as a small settlement and spread out, or that commerce activities prefer to be close to primary roads. But why does it happen this way and not otherwise? If we do not have an explanation, we cannot "translate" the behavior into a computer model, identifying the "active urban principles" from the rest that are not relevant to the urban phenomena of interest. It can be asserted that everything is related and influences everything else, but in modeling, only the factors that are significantly responsible for the behavior of interest are considered. This becomes an exercise of pondering and approximations. In the case of explaining why a settlement grows mainly from the inside out, it is observed that small increments of growth occur in the vicinity of previously urbanized areas. This is the behavior. It is useful, but the most important is to identify its causes, which will subsequently be encoded in a computer program.

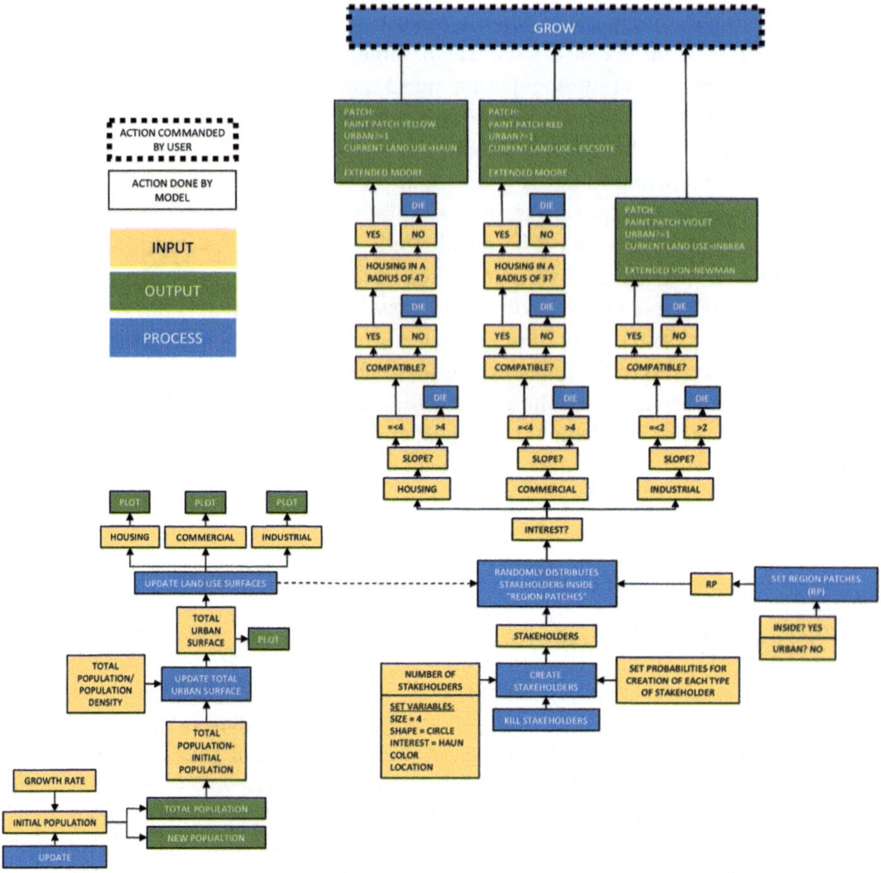

Fig. 16 Example diagram of an urban growth model

The causes can be that people prefer to be in a formal, service-provided area with available water, electricity, drainage and road services instead of residing in a squatter. When this urbanization behavior is coded in a model, such conditions are sought in the territory by its people-representing agents.

Once the causes of behavior have been established, the next step will be to formalize them in a computational language, and a recommendable previous step is to make a diagram that visually establishes the order, hierarchy, and relations between causes (Fig. 16). The diagram is not only a compass that keeps the modeler on path when coding, but by drawing the diagram orders the ideas and even helps to identify errors or causes of unexpected outputs of a model that cannot be identified otherwise, or at least not as easy.

Finally, some words of caution. If a theory manages to reproduce an observed behavior in a computer model, it can be said that, adapting Epstein's phrase, "If you could grow it, you found one way to explain it." And only one, from an unknown number of possible explanations. For example, there is a well-known computational model that reproduces bird flocking with only three rules [148], but it has never been claimed that these are the same ones that birds follow in real life. In the same way, we must be cautious with theories, since we are facing with something similar with what in systems theory is called equifinality, or the possibility of reaching the same final state from different initial conditions. Given this situation, it is best to stick to what was said by statistician George Box: "all models are wrong, but some are useful" [149].

4.3.1 Theory Formalization in a Programming Language

Having the theory, now it must be formalized in a computational language. In this step, a programming specialist should alleviate the workload and with which there would be close communication. If this is not possible, learning a programming language is the other option, which is a task that demands discipline and dedication, particularly for those who, in the planning field, is not a habitual skill. For this scenario, here are some recommendations that will facilitate this task:

- *Abstraction of the phenomenon of interest.* Part of the process of developing our models is to observe the real-life target to be reproduced and abstract it to a level in which similar behaviors can be found in other models. For example, when spatially studying cities, we can talk about agglomeration, segregation, and growth processes, so we would be looking for models that, although not of cities, reproduce these processes.
- *Search for similar models.* There is no exclusive programming language for cities, but there are models and code fragments that reproduce their behaviors. Therefore, instead of wanting to learn all the possibilities of a programming language, the focus should be on identifying and learning from those models that similarly reproduce behaviors of interest, identify the code and adapt it to what is being modeled (Table 4). And searching does not have to be literal. The best models are those that can be applied to different fields. The fact that a code reproduces the behavior of a flock of birds does not mean that it is not useful to reproduce crowd movement.
- *Identifying variables that can be modified from a base model.* As already said, the Game of Life is a common base for more elaborate models, but it is also good for programming learning. Its code is not extensive; it is easy to read and the resulting behaviors are varied. Once the commands and structure of a chosen model code have been understood, a good practice for those who wish to learn to program is to make changes on their own. It is recommended first to identify the variables that can be modified and the effects they can cause, and only then try to obtain a specific behavior, not the other way around.
- *Drawing a flow diagram of the model's sequence.* A flow diagram will be the visual representation of the code to be written. Visualizing the code clarifies a models sequence, helps identify and control the agents involved, their states, the rules of

Table 4 Sample of Netlogo models as of version 6.0.3 that may be relevant for urban model building

Model name	Description	Interest
Hex cell aggregation	Growth model. It takes place on a two-dimensional hexagonal grid of cells. Cells can either be alive or dead. Various growth patterns result, depending on the exact rules governing growth	Aggregation Cellular automata
Erosion	Simulation of soil erosion by water. The user is presented with an empty terrain. Rain falls on the terrain and starts to flow downhill. As it flows, it erodes the terrain below. The patterns of water flow change as the terrain is reshaped by erosion. Eventually, a river system emerges	Random terrain generation
Grand Canyon	Simulation of rainfall on a patch of terrain on the eastern end of the Grand Canyon	Elevation model generation
Segregation	Behavior of two types of agents in a neighborhood. The red agents and green agents get along with one another. But each agent wants to make sure that it lives near some of "its own." That is, each red agent wants to live near at least some red agents, and each green agent wants to live near at least some green agents. The simulation shows how these individual preferences ripple through the neighborhood, leading to large-scale patterns	Aggregation Cellular automata
Voting	Cellular automaton that simulates voting distribution by having each patch take a "vote" of its eight surrounding neighbors, then perhaps change its own vote according to the outcome	Aggregation Cellular automata Cell value according to neighbors
Wealth distribution	Wealth distribution model	Terrain generation with values variable by patch
Artificial Anasazi	Population dynamics simulation in the long house valley in Arizona between 800 and 1400	Territorial information layers display management
Economic disparity	Residential land-usage patterns from an economic perspective, using the socio-economic status of the agents to determine their preferences for choosing a location to live	Terrain generation with values variable by patch Territorial information layers display management Variables behavior affects patches values

(continued)

Table 4 (continued)

Model name	Description	Interest
Sprawl effect	City growth and how it leads to urban sprawl and the problems connected with it (e.g. leapfrogging)	Topography of attraction generation
Structure from randomness 2	Each grid cell is initialized to have a potential value of either 1 or -1, at random Each tick of the model, the cell takes on a new potential value, which is the average of the potential values from its Von Neumann neighborhood (that is, itself and its 4 neighbors) from the previous step, plus a random number, either 1 or -1	Aggregation Cellular automata Cell value according to neighbors
CSV example	NetLogo's csv extension to load data from and write data to data to CSV files	Data loading from external file
GIS general examples	GIS NetLogo extension functionality example	SHP layer loading (points, vectors and raster) Patch properties according to SHP data

interaction and the environment in which they operate. Diagramming a model also shows that it is composed of more parts and interactions than we might think at first.

- *Writing a literal description of the model.* Writing the model is a process that goes hand in hand with drawing a flow diagram drawing. It's done by answering the following questions:

  – What part of the phenomenon is going to be modeled?
  – What are the types of agents involved in this phenomenon?
  – What actions or behaviors can these agents take? (with each other and with the environment)
  – In what type of environment do these agents operate?
  – If the phenomenon is defined as discrete time steps, what events could happen at each step and in what order?
  – What is expected to be observed from the model?

  Answering these questions gives feedback to elaborate a protocol called ODD (Overview, Design Concepts, Details) [150, 151], which is a guide that helps describe, understand and communicate Agent-Based Models.
- *Build with blocks.* For those who are new to programming, it is a challenge to fluently write all the code of a model. Based on a previously prepared flow diagram, it is convenient to divide the model into blocks of simple, controllable and measurable actions. A block can be the action of painting lands of a particular color,

creating agents, the interaction between them or with their environment, and so on. The key words in this process are *control* and *expected result*, both in what we want to create and the interactions we want to code. Executing these blocks independently, in controlled conditions, and in such scenarios that we can know in advance what the result should be is a verification test (internal consistency) of the model.

- *Identify the code structure.* Once models with behaviors we are seeking to reproduce are identified, what follows is to understand their code structure. Although the code of a model is displayed fluently, its execution is not given in that order. The instructions are provided by "packages," and they are related to each other in a precise way.
- *Start with a simple model.* Once our modeling blocks have been identified, the most practical thing is to start with one of them, verify their effectiveness, and add more blocks until the desired model is achieved. The keyword here is *manageability*.
- *Build a snippets library.* A snippet is a small piece of reusable code. For this reason, we should review as many models as possible in search of behaviors that we know will be useful for our models, identify their code, isolate and store them in a snippets library. The treatment that policy makers and computer modelers give to their work products is different. Unlike the former, a computer modeler must keep a controlled record of previous model versions, since they can add value to later versions. In a circular process of experimentation, analysis and new experimentation where programs are created to emulate the behavior of a system, a "box of nuts and bolts" (or fragments of code) is accumulated over time. Although they will not fit in all the needs of a current model, they will surely serve purposes for later ones.
- *Keep a record of model versions.* The final model with which the experiments will be made is not the result of one, but several evolving models. As functionalities are added and understanding of the system progresses, it may happen that what was once useful or interesting later becomes irrelevant. The research question can even change due to two intrinsic characteristics of modeling: experimentation and discovery. This process is circular, so it is not uncommon to go back a few steps to retake previously eliminated elements. For this reason, it is imperative to keep an orderly record of changes, differentiated by ascending numbers of model versions, where minor revisions are represented by decimal increments to the current version number, or adding whole numbers representing a substantial model change. Besides a version log, it is also useful to write a to-do list in which the tasks to be done is written.
- *Interact with online programming communities.* Support in learning and the model building process can be obtained in various online communities, whose ultimate goal is to help among its members. The advantage they present over written material is that they manage the most up to the moment information and that specific questions can be asked with a high probability that they will be answered at some point.

Although programming is initially overwhelming, fortunately, there is literature regarding the development of Agent-Based Models with enough practical examples that facilitate learning. Such titles are, among others, Spatial Simulation: Exploring Pattern and Process [71], An Introduction to Agent-Based Modeling [123], Introduction to the Modeling and Analysis of Complex Systems [152], Agent-Based and Individual-Based Modeling [124], Netlogo: A Modeling Tool [153] and Fundamentals of Multiagent Systems [154].

Here, too, some words of caution. Programming the imitation of behavior is not the same as programming the structural causes that generate it. Returning to the bird flocking example, the modeler did not program in advance the path that each bird had to follow. What he programmed was the obedience of all the birds to three rules of behavior, and leave the birds to fly by themselves. Although none of the rules explicitly dictates "form groups that fly in a sinuous way," the result is this behavior.

4.4 Model Execution

The modeling process offers the freedom to experiment, something normally prohibitive in traditional urban planning processes, in which the final proposal is the result of assumptions detached from reasoning that could be enriched if the interactions were understood. For this reason, modeling allows the exploration of alternatives. Although this demands more time, it offers the possibility of discovery. This exploratory feature has two sides: it can show that the assumptions are not such or what was expected as a result does not finally arrive. It can also show new lines of research that are only discovered after observation and analysis of concrete results.

4.5 Output Analysis and Representation

The modeling process also requires a variety of tools and methods for processing, analysis, and visualization of outputs. For data processing, it is necessary to have an organized workflow that indicates which, how, when and what tools will be used. This is software that ranges from Geographical Information Systems, spreadsheets, programming, to image editing and bibliographic database administration. It is also necessary to know what are the existing methods to analyze, represent and visualize a models output, which becomes particularly complicated because it is not static but changes in a quantitative and spatial way [155]. Visualization on paper is currently a limiting factor, and for this reason, its representation is now explored through digital and interactive media [156].

Finally, in spite of the advantages that computational modeling of urban phenomena could offer, a series of challenges are present that hinder its wide implementation and practice. One is the difficulty that a single professional can effectively perform all the necessary tasks that the modeling process demands. Specialists in geographic information systems, programmers and planners are just some of the members of what should be a multidisciplinary team that would be complemented with other profiles according to the needs of the model, such as sociologists and economists. Another significant challenge is the availability of spatial data that's properly digitized, updated, ordered and verified.

4.6 Results Presentation

Once the phenomenon to be modeled has been chosen, a theory has been elaborated about why it happens, identified the active components of this theory, established their relationships and interactions, translate them into a computational model and executed the model; what follows is to analyze and present the results, aimed at interested audience such as the scientific community and, in the context of public policies, to stakeholders and decision makers.

Taking into account all the phases of the modeling process, it is clear that the development of computer models is an elaborated task that takes place in the context of a nascent discipline and used by many differentiated disciplines. For this reason, there is still no standardized protocol to present models in such a way that they can be adequately communicated. Published descriptions of models are often difficult to read, incomplete, ambiguous and therefore less accessible than other analytical models, not giving readers the ability to better judge the contribution of simulation studies. If models are not structured and presented in an understandable manner, with enough information and transparency in how they work, they cannot be adequately reproduced, slowing down knowledge creation and leading to duplication of effort in research [150, 157].

As a way to overcome these obstacles, group efforts have been made to create guidelines to be used as a standard to structure and communicate models. One of them is the ODD (Overview, Design concepts, Details) protocol [150, 151]. It is composed of three parts: a general context and information section (Overview), a description of the general concepts of the model (Design Concepts) and technical details (Details) (Table 5). Although this protocol arises from an ecological perspective, it has been transferred to models of human interaction. Nevertheless, an adaptation called ODD+D has been made for a more appropriate implementation when applied to agent-based models related to human activities [158]. By its nature, this protocol adds additional elements which focus on characterizing human decision making, particularly in empirical, theoretical and decision-making aspects at the individual level. Another protocol is the recently published STRESS (Strengthening the Reporting of Empirical Simulation Studies) guidelines [157], which can be applied not only to agent-based models but also to discrete-event simulation and system dynamics (Table 6).

Something to note about published works regarding the importance of these protocols, is that their focus is on the utility they have as a facilitator of communication of models, but they do not emphasize about the usefulness they also have as a tool

Table 5 Elements of the ODD protocol, grouped into the three parts: overview, design concepts, and details [150, 151]

Overview	Purpose
	Entities, state variables and scales
	Process overview and scheduling
Design concepts	Design concepts
Details	Initialization
	Input data
	Submodels

Table 6 Elements of the STRESS guidelines [157]

Section	Checklist item
1. Objectives	Purpose of the model
	Model outputs
	Experimentation aims
2. Logic	Base model overview diagram
	Base model logic
	Scenario logic
	Algorithms
	Components
3. Data	Data sources
	Input parameters
	Pre-processing
	Assumptions
4. Experimentation	Initialization
	Run length
	Estimation approach
5. Implementation	Software or programming language
	Random sampling
	Model execution
	System specification
6. Code access	Computer model sharing statement

to guide the whole modeling process itself, from its conception to its application in a specific experiment. Just as a graphical diagram of the code to be implemented helps the programmer to be focused on their work, these protocols can have the same function: they help maintain an orderly and structured modeling process, from the model's conception to its final application in a particular experiment. Thus, a protocol can function as the guiding element that provides control and clarity, developed from the initial stages of modeling, not at the end as a mere report, as is sometimes understood when reading about these guides. If they are to be taken as tools of constant reference in the modeling process, this will entail the need to keep a record of how their contents are changing as the models they refer to change along the creation process.

4.6.1 Communicating Models to a Wider Audience

Although these protocols can fulfill their function before a specialized reader, if we want to have an impact on decision-making, they must also be conceived for a broader audience, composed of heterogeneous profiles of specialization, available listening time and even current attention span. It is better to work on audience-tailored agendas when possible, but in broad contexts such as public meetings, it can be a difficult task. It must be recognized that scientific, citizen, political and business languages vary and that they share sympathies and adversities that are irrelevant to the model but may affect its perception even before it is presented. For this reason, it is important to show only relevant information and in a way that does not give rise to ambiguous interpretation due to incomplete information or scientific jargon. For this, a guide for model communication aimed at the non-specialized audience is suggested (Table 7), along with a series of considerations that must be taken into account in advance before a presentation according to questions that may arise as part of the final policy

Table 7 Presentation guide to non-specialized audience in the context of public policy decision making, based on [157]

Section	Checklist item	Considerations
1. Objectives	Purpose of the model	Action oriented Primarily concerned with social action Government-oriented problem solving Identify audience-relatable model targets (geographic area, social groups, current government challenges, public demands)
2. Logic	Model overview diagram	Less interested in process, more interested in outcomes Less interested in details, more interested in generalities More graphical than text oriented
3. Results	Output presentation	Less interested in explanations, more interested in description, prediction and application Results must be generalizable to a wider population Use of succinct, jargon-free language and of an executive summary
4. Real-life implementation	Translation from model outputs to legal norms	Identify which laws, ordinances, programs and other institutional instruments must be considered/adapted for implementation Political implications of implementation Timeframe implementation Human resources and funding needs

implementation process. Although many of them may not be relevant to the model itself, they are relevant to stakeholders and for the following phases of public policy implementation that could result from the use of agent-based models.

References

1. Gutierrez Chaparro, J.J.: Planeacion urbana en México: un analysis critico sobre su proceso de evolucion. Red Rev Científicas América Lat el Caribe, España y Port **12**, 52–63 (2009)
2. Portugali, J., Meyer, H., Stolk, E., Tan, E.: Complexity Theories of Cities have Come of Age, 1st edn. Springer (2012)
3. Lynch, K.: A Theory of Good City Form. MIT Press (1981)
4. Bettencourt, L.M.A.: The origins of scaling in cities. Science **340**(80), 1438–1441 (2013). https://doi.org/10.1126/science.1235823
5. Marshall, S.: City planning, design and evolution concepts (2009)
6. Sandoval-Félix, J., Castañón-Puga, M., Gaxiola-Pacheco, C., Suarez, E.: Identifying clusters of complex urban–rural issues as part of policy making process using a network analysis approach: a case study in Bahía de Los Ángeles, México. Sustainability **9**, 1059 (2017). https://doi.org/10.3390/su9061059
7. OECD, Global Science Forum: Applications of complexity science for public policy: new tools for finding unanticipated consequences and unrealized opportunities. Erice, Sicily (2009)
8. Alves, B., Sakowski, P., Tovolli, M.: Modeling Complex Systems for Public Policies, 1st edn. Institute for Applied Economic Research, Brasília (2015)
9. Ball, P.: Why Society is a Complex Matter: Meeting Twenty-First Century Challenges with a New Kind of Science, 1st edn. Springer (2012)
10. IMIP: Estudio geoestadístico del fenómeno delictivo. Ensenada, México (2014)
11. IMIP: Programa parcial de desarrollo urbano Maneadero-Punta Banda, Ensenada, B.C. Ensenada, México (2012)
12. Sandoval-Félix, J.: Adecuacion y reprogramacion del software urban growth simulator para su utilizacion en los procesos de planeacion de la ciudad de Ensenada, B.C., Mexico. Archit. City Environ. 91–110 (2011)
13. Weaver, W.: Science and complexity. Am. Sci. **36**, 536–544 (1948). https://doi.org/10.1007/978-1-4899-0718-9_30
14. Filatova, T.: Empirical agent-based land market: integrating adaptive economic behavior in urban land-use models. Comput. Environ. Urban Syst. **54**, 397–413 (2015). https://doi.org/10.1016/j.compenvurbsys.2014.06.007
15. Edmonds, B., Meyer, R.: Simulating Social Complexity: A Handbook, 1st edn. Springer, Manchester (2013)
16. Weiler, R., Engelbrecht, J.: The new sciences of networks & complexity: a short introduction. Cadmus J. **2**, 131–141 (2013)
17. Beheshti, R., Sukthankar, G.: A hybrid modeling approach for parking and traffic prediction in urban simulations. AI Soc. 1–12 (2014)
18. Easley, D., Kleinberg, J.: Networks, crowds, and markets: reasoning about a highly connected world. Science **81**(80), 744 (2010). https://doi.org/10.1017/CBO9780511761942
19. Morcol, G.: Urban sprawl and public policy: a complexity theory perspective. Emerg. Complex Organ. **14**, 1–16 (2012)
20. Batty, M.: Cities as complex systems: scaling, interactions, networks, dynamics and urban morphologies. London (2008)
21. Read, S.: Meaning and material: phenomenology, complexity, science and adjacent possible cities. In: Complexity Theories of Cities have Come of Age, pp. 105–127 (2012)
22. Mitchell, M.: Complexity: A Guided Tour, 1st edn. Oxford University Press, New York (2009)

23. Goldstein, J.: Emergence as a construct: history and issues. Emergence **1**, 49–72 (1999). https://doi.org/10.1207/s15327000em0101_4
24. Gilbert, N., Troitzcsch, K.G.: Simulation for the Social Scientist, 2nd edn. Open University Press, Glasgow (2005)
25. Anderson, P.W.: More is different. Science **177**, 393–396 (1972). https://doi.org/10.1126/science.177.4047.393
26. Batty, M., Marshall, S.: The origins of complexity theory in cities and planning. In: Complexity Theories of Cities have Come of Age, pp. 21–45 (2012)
27. Furtado, B.A.: Policy space: agent based modeling, 1st edn. Institute for Applied Economic Research—IPEA, Rio de Janeiro (2018)
28. Page, S.E.: Preface. In: Alves, B., Sakowski, P., Tovolli, M. (eds.) Modeling Complex Systems for Public Policies, 1st edn, pp. 9–14. Institute for Applied Economic Research, Brasília (2015)
29. Gentile, J.E., Glazner, C., Koehler, M.: Simulation models for public policy. In: Alves, B., Sakowski, P., Tovolli, M. (eds.) Modeling Complex Systems for Public Policies, 1st edn, pp. 73–83. Institute for Applied Economic Research, Brasília (2015)
30. Bettencourt, L.M.A.: Cities as complex systems. In: Alves, B., Sakowski, P., Tovolli, M. (eds.) Modeling Complex Systems for Public Policies, pp. 217–236. Institute for Applied Economic Research, Brasília (2015)
31. Rand, W.: Complex systems: concepts, literature, possibilities and limitations. In: Alves, B., Sakowski, P., Tovolli, M. (eds.) Modeling Complex Systems for Public Policies, pp. 37–54. Institute for Applied Economic Research, Brasília (2015)
32. Walloth, C., Gurr, J.M., Schmidt, J.A.: Understanding Complex Urban Systems: Multidisciplinary Approaches to Modeling, 1st edn. Springer, Switzerland (2014)
33. Cairney, P.: An advocacy coalition framework of policy change and the role of policy-oriented learning therein. In: Page, E., Balla, S., Lodge, M. (eds.) Oxford Handbook of the Classics of Public Policy and Administration. Oxford, p. 13 (2012)
34. Bar-Yam, Y.: Complexity theory in applied policy worldwide. In: Alves, B., Sakowski, P., Tovolli, M. (eds.) Modeling Complex Systems for Public Policies, 1st edn, pp. 239–259. Institute for Applied Economic Research, Brasília (2015)
35. Anderson, E.G.: An initial simulation model for aiding policy analysis in urban insurgencies. In: Proceedings—Winter Simulation Conference, pp. 1168–1176 (2007)
36. Widener, M.J., Metcalf, S.S., Bar-Yam, Y.: Agent-based modeling of policies to improve urban food access for low-income populations. Appl. Geogr. **40**, 1–10 (2013). https://doi.org/10.1016/j.apgeog.2013.01.003
37. Gomes, O.: The economy as a complex object. In: Alves, B., Sakowski, P., Tovolli, M. (eds.) Modeling Complex Systems for Public Policies, pp. 169–190. Institute for Applied Economic Research, Brasília (2015)
38. Kenbeek, S.D., Bone, C., Moseley, C.: A network modeling approach to policy implementation in natural resource management agencies. Comput. Environ. Urban Syst. **57**, 155–177 (2016). https://doi.org/10.1016/j.compenvurbsys.2016.02.003
39. Yarime, M., Kharrazi, A.: Understanding the environment as a complex, dynamic natural-social system: opportunities and challenges in public policies for promoting global sustainability. In: Alves, B., Sakowski, P., Tovolli, M. (eds.) Modeling Complex Systems for Public Policies. Institute for Applied Economic Research, pp. 127–140 (2015)
40. Lee, J.: Development of a community-accessible urban sprawl impact assessment system for Northeast Ohio (2001)
41. Noth, M., Borning, A., Waddell, P.: An extensible, modular architecture for simulating urban development, transportation, and environmental impacts. Comput. Environ. Urban Syst. **27**, 181–203 (2003). https://doi.org/10.1016/S0198-9715(01)00030-8
42. Johnson, J.R., Nowak, A., Ormerod, P., et al.: Non-equilibrium social science and policy introduction and essays on new and changing paradigms in socio-economic thinking, 1st edn. Springer (2017)
43. Tessone, C.J.: The complex nature of social systems. In: Alves, B., Sakowski, P., Tovolli, M. (eds.) Modeling Complex Systems for Public Policies, pp. 141–168. Institute for Applied Economic Research, Brasília (2015)

44. Haghshenas, H., Vaziri, M., Gholamialam, A.: Evaluation of sustainable policy in urban transportation using system dynamics and world cities data: a case study in Isfahan. Cities **45**, 104–115 (2015). https://doi.org/10.1016/j.cities.2014.11.003
45. Roumboutsos, A., Kapros, S.: A game theory approach to urban public transport integration policy. Transp. Policy **15**, 209–215 (2008). https://doi.org/10.1016/j.tranpol.2008.05.001
46. Almeida, A.: Overcoming chaos: legislatures as complex adaptive systems. In: Alves, B., Sakowski, P., Tovolli, M. (eds.) Modeling Complex Systems for Public Policies, pp. 337–362. Institute for Applied Economic Research, Brasília (2015)
47. Jacobson, M.J.: Education as a complex system: implications for educational research and policy. In: Modeling Complex Systems for Public Policies. Institute for Applied Economic Research, pp. 301–314 (2015)
48. Ahrweiler, P., Schilperoord, M., Pyka, A., Gilbert, N.: Modelling research policy: ex-ante evaluation of complex policy instruments. Jasss **18** (2015). https://doi.org/10.18564/jasss.2927
49. Sandoval-Félix, J., Castañón-Puga, M.: Using an urban growth simulator for Ensenada city public policy analysis. In: Computational Social Science Society of the Americas (CSSSA). Santa Fe, pp. 1–12 (2015)
50. Chettiparamb, A.: Complexity theory and planning: examining “fractals” for organising policy domains in planning practice. Plan Theory **13**, 5–25 (2013). https://doi.org/10.1177/1473095212469868
51. Meek, J.: Complexity theory for public administration and policy. Emerg. Complex Organ. **12** (2010)
52. Portugali, J.: Complexity theories of cities: achievements, criticism and potentials. In: Complex Theory for Cities have Come Age, pp. 1–14 (2009). https://doi.org/10.1007/978-3-642-24544-2
53. Jacobs, J.: The Death and Life of Great American Cities. Random House, New York (1961)
54. Batty, M.: Complexity in city systems: understanding, evolution, and design. London (2007)
55. Portugali, J.: Complexity, Cognition and the City, 1st edn. Springer, Tel Aviv (2011)
56. Atun, F.: Understanding effects of complexity in cities during disasters. In: Walloth, C., Gurr, J.M., Schmidt, J.A. (eds.) Understanding Complex Urban Systems: Multidisciplinary Approaches to Modeling. Springer, pp. 51–65 (2014)
57. Murgante, B., Borruso, G., Lapucci, A.: Geocomputation and Urban Planning (2009)
58. Alberti, M.: Advances in Urban Ecology. Springer, New York (2008)
59. Roo, G. De Silva, E.A.: A Planner’s Encounter with Complexity. ASHGATE (2010)
60. Xie, F., Levinson, D.M.: Evolving Transportation Networks (2011)
61. Heppenstall, A., Crooks, A., See, L., Batty, M.: Agent-Based Models of Geographical Systems, 1st edn. Springer (2012)
62. Basse, R.M., Omrani, H., Charif, O., et al.: Land use changes modelling using advanced methods: cellular automata and artificial neural networks. The spatial and explicit representation of land cover dynamics at the cross-border region scale. Appl. Geogr. **53**, 160–171 (2014). https://doi.org/10.1016/j.apgeog.2014.06.016
63. Liu, Y., He, Q., Tan, R., et al.: Modeling different urban growth patterns based on the evolution of urban form: a case study from Huangpi, Central China. Appl. Geogr. **66**, 109–118 (2016). https://doi.org/10.1016/j.apgeog.2015.11.012
64. Wissen Hayek, U., Efthymiou, D., Farooq, B., et al.: Quality of urban patterns: spatially explicit evidence for multiple scales. Landsc. Urban Plan. **142**, 47–62 (2015). https://doi.org/10.1016/j.landurbplan.2015.05.010
65. Peng, J., Liu, Y., Wu, J., et al.: Landscape and urban planning linking ecosystem services and landscape patterns to assess urban ecosystem health: a case study in Shenzhen City, China. Landsc. Urban Plan. **143**, 56–68 (2015). https://doi.org/10.1016/j.landurbplan.2015.06.007
66. Rezayan, H., Delavar, M.R., Frank, A.U., Mansouri, A.: Spatial rules that generate urban patterns: emergence of the power law in the distribution of axial line length. Int. J. Appl. Earth Obs. Geoinf. **12**, 317–330 (2010). https://doi.org/10.1016/j.jag.2010.04.003

67. Frantál, B., Greer-Wootten, B., Klusáček, P., et al.: Exploring spatial patterns of urban brownfields regeneration: the case of Brno, Czech Republic. Cities **44**, 9–18 (2015). https://doi.org/10.1016/j.cities.2014.12.007
68. Holt, A.R., Mears, M., Maltby, L., Warren, P.: Understanding spatial patterns in the production of multiple urban ecosystem services. Ecosyst. Serv. **16**, 33–46 (2015). https://doi.org/10.1016/j.ecoser.2015.08.007
69. Gupta, R.: The pattern of urban land-use changes: a case study of the indian cities. Environ. Urban Asia **5**, 83–104 (2014). https://doi.org/10.1177/0975425314521539
70. Decraene, J., Monterola. C., Lee, G.K.K., et al.: The emergence of urban land use patterns driven by dispersion and aggregation mechanisms. PLoS One **8** (2013). https://doi.org/10.1371/journal.pone.0080309
71. O'Sullivan, D., Perry, G.L.W.: Spatial Simulation: Exploring Pattern and Process (2013)
72. Hossain, S., Scholz, W., Baumgart, S.: Translation of urban planning models: planning principles, procedural elements and institutional settings. Habitat. Int. **48**, 140–148 (2015). https://doi.org/10.1016/j.habitatint.2015.03.006
73. Aguilera, A., Ugalde, E.: A spatially extended model for residential segregation. Discret. Dyn. Nat. Soc. (2007). https://doi.org/10.1155/2007/48589
74. Marull, J., Font, C., Boix, R.: Modelling urban networks at mega-regional scale: are increasingly complex urban systems sustainable? Land Use Policy **43**, 15–27 (2015). https://doi.org/10.1016/j.landusepol.2014.10.014
75. Li, M.: Social network and social capital in leadership and management research: a review of causal methods. Leadersh Q **24**, 638–665 (2013). https://doi.org/10.1016/j.leaqua.2013.04.005
76. Hansen, D., Shneiderman, B., Smith, M.A.: Analyzing social media networks with NodeXL: insightsm a connected world, 1st edn. Elsevier, Burlington (2011)
77. Wakita, K., Tsurumi, T.: Finding community structure in mega-scale social networks (2007). arXiv:cs/0702048[csCY]1275, https://doi.org/10.1145/1242572.1242805
78. Clauset, A., Newman, M.E.J., Moore, C.: Finding community structure in very large networks. Phys. Rev. E **70**, 66111 (2004). https://doi.org/10.1103/PhysRevE.70.066111
79. Hélie, M.: Conceptualizing the principles of emergent urbanism. Int. J. Archit. Res. **3**, 75–91 (2009)
80. Batty, M.: Complexity and emergence in city systems: implications for urban planning **10**, 15–32 (2009)
81. Roy, D., Lees, M.H., Palavalli, B., et al.: The emergence of slums: a contemporary view on simulation models. Environ. Model Softw. **59**, 76–90 (2014)
82. Batty, M.: The emergence of cities: complexity and urban dynamics. London (2003)
83. Batty, M.: Cities and Complexity: Understanding Cities with Cellular Automata, Agent-Based Models, and Fractals, 1st edn. MIT Press, Cambridge, Massachusetts (2007)
84. Solé, R.V.: Phase Transitions, 1st edn. Princeton University Press, New Jersey (2011)
85. Wilson, A.: The Science of Cities and Regions: Lectures on Mathematical Model Design. Springer, London (2012)
86. Aljoufie, M., Zuidgeest, M., Brussel, M., et al.: A cellular automata-based land use and transport interaction model applied to Jeddah, Saudi Arabia. Landsc. Urban Plan. **112**, 89–99 (2013). https://doi.org/10.1016/j.landurbplan.2013.01.003
87. Gatta, V., Marcucci, E.: Behavioural implications of non-linear effects on urban freight transport policies: the case of retailers and transport providers in Rome. Case Stud. Transp. Policy (2015). https://doi.org/10.1016/j.cstp.2015.08.001
88. Crooks, A., Castle, C., Batty, M.: Key challenges in agent-based modelling for geo-spatial simulation. Comput. Environ. Urban Syst. **32**, 417–430 (2008). https://doi.org/10.1016/j.compenvurbsys.2008.09.004
89. Gaube, V., Remesch, A.: Impact of urban planning on household's residential decisions: an agent-based simulation model for Vienna. Environ. Model Softw. **45**, 92–103 (2013). https://doi.org/10.1016/j.envsoft.2012.11.012

90. Jordan, R., Birkin, M., Evans, A.: Agent-based modelling of residential mobility, housing choice and regeneration. In: Heppenstall, A.J., Crooks, A.T., See, L.M., Batty, M. (eds.) Agent-Based Models of Geographical Systems, pp. 511–524. Springer (2013)
91. Garrocho Rangel, C., Chavez, T., Alvarez, J.A.: La dimension espacial de la competencia comercial. El Colegio Mexiquense (2002)
92. Mahabir, R., Crooks, A., Croitoru, A., et al.: The study of slums as social and physical constructs: challenges and emerging research opportunities. Reg. Stud. Reg. Sci. **3**, 737–757 (2016). https://doi.org/10.1080/21681376.2016.1229130
93. Patel, A.V.: Slumulation: an integrated simulation framework to explore spatio-temporal dynamic of slum formation in Ahmedabad, India (2012)
94. Barros, J., Sobreira, F.: City of slums: self-organisation across scales (2002)
95. Deep, S., Saklani, A.: Urban sprawl modeling using cellular automata. Egypt. J. Remote Sens. Space. Sci. **17**, 179–187 (2014). https://doi.org/10.1016/j.ejrs.2014.07.001
96. Lagarias, A.: Urban sprawl simulation linking macro-scale processes to micro-dynamics through cellular automata, an application in Thessaloniki, Greece. Appl. Geogr. **34**, 146–160 (2012). https://doi.org/10.1016/j.apgeog.2011.10.018
97. Torrens, P., Kevrekidis, I., Ghanem, R., Zou, Y.: Simple urban simulation atop complicated models: multi-scale equation-free computing of sprawl using geographic automata. Entropy **15**, 2606–2634 (2013). https://doi.org/10.3390/e15072606
98. Allen, P.: Spatial planning in a complex unpredictable world of change: towards a proactive co-evolutionary type of planning within the Eurodelta, 1st edn. Coöperatie In Planning UA, Groningen (2016)
99. Zhou, M.: An interval fuzzy chance-constrained programming model for sustainable urban land-use planning and land use policy analysis. Land Use Policy **42**, 479–491 (2015). https://doi.org/10.1016/j.landusepol.2014.09.002
100. Sakieh, Y., Salmanmahiny, A., Jafarnezhad, J., et al.: Evaluating the strategy of decentralized urban land-use planning in a developing region. Land Use Policy **48**, 534–551 (2015). https://doi.org/10.1016/j.landusepol.2015.07.004
101. Crooks, A., Patel, A., Wise, S.: Multi-agent systems for urban planning. In: Technologies for Urban and Spatial Planning: Virtual Cities and Territories, pp. 29–56 (2014)
102. De, R.G., Hillier, J., Van, W.J.: Complexity and Planning: Systems, Assemblages and Simulations (New Directions in Planning Theory), 1st edn. Ashgate (2012)
103. De, R.G.: Spatial planning, complexity and a world 'out of equilibrium': outline of a non-linear approach to planning. In: Complexity and Planning Systems: Assemblages and Simulations, pp. 129–165 (2011)
104. Porta, J., Parapar, J., Doallo, R., et al.: High performance genetic algorithm for land use planning. Comput. Environ. Urban Syst. **37**, 45–58 (2013). https://doi.org/10.1016/j.compenvurbsys.2012.05.003
105. Gar-On, Y., Xia, L.: Urban simulation using neural networks and cellular automata for land use planning. In: Symposium on Geospatial Theory, Processing and Applications. Otawwa (2002)
106. Zia, K., Riener, A., Farrahi, K., Ferscha, A.: A new opportunity to urban evacuation analysis: very large scale simulations of social agent systems in repast HPC. In: Proceedings—2012 ACM/IEEE/SCS 26th Workshop on Principles of Advanced and Distributed Simulation, PADS 2012, pp. 233–242 (2012)
107. Allen, P.M.: Cities: the visible expression of co-evolving complexity. In: Complexity Theories of Cities have Come of Age, pp. 67–89 (2012)
108. Pumain, D.: Urban systems dynamics, urban growth and scaling laws: the question of ergodicity. In: Complexity Theories of Cities have Come of Age, p. 91 (2012). https://doi.org/10.1007/978-3-642-24544-2_6
109. Hillier, B.: The genetic code for cities: is it simpler than we think? In: Complexity Theories of Cities have Come of Age, pp. 129–152 (2012). https://doi.org/10.1007/978-3-642-24544-2
110. Johnson, J.: Cities: systems of systems of systems. Sustain. Dev. 153–172 (2012)
111. Daoudi, M.: The Experimental City, 1st edn. Routledge, New York (2016)

112. Vigil, G.M.: Diseño urbano: teoría y método, 2nd edn. Ediciones Universidad Católica de Chile, México, D.F. (2000)
113. OECD Global Science Forum: Effective modelling of urban systems to address the challenges of climate change and sustainability. In: OECD Global Science Forum, p. 99 (2011)
114. Castelliani, B., Hafferty, F.: Sociology and Complexity Science. A New Field of Inquiry. Springer, Heidelberg (2009)
115. Furtado, B.A., Sakowski, P.A.M., Tóvolli, M.H.: A complexity approach for public policies. Inst Pesqui Econômica Apl (2015)
116. Fang, S., Gertner, G.Z., Sun, Z., Anderson, A.A.: The impact of interactions in spatial simulation of the dynamics of urban sprawl. Landsc. Urban Plan. **73**, 294–306 (2005). https://doi.org/10.1016/j.landurbplan.2004.08.006
117. Sudhira, H.S., Ramachandra, T.V., Jagadish, K.S.: Urban sprawl: metrics, dynamics and modelling using GIS. Appl. Earth Obs. Geoinf. **5**, 29–39 (2004). https://doi.org/10.1016/j.jag.2003.08.002
118. Axelrod, R.: Advancing the art of simulation in the social sciences. J. Jpn. Int. Econ. **12**, 16–22 (2003). https://doi.org/10.1002/(SICI)1099-0526(199711/12)3:2%3c16:AID-CPLX4%3e3.0.CO;2-K
119. Conte, R., Gilbert, N., Bonelli, G., et al.: Manifesto of computational social science. Eur. Phys. J. Spec. Top. **214**, 325–346 (2012). https://doi.org/10.1140/epjst/e2012-01697-8
120. Cioffi-Revilla, C.: Computational social science. Wiley Interdiscip. Rev. Comput. Stat. **2**, 259–271 (2010). https://doi.org/10.1002/wics.95
121. Gries, D., Schneider, F.B., Cioffi-Revilla, C.: Introduction to Computational Social Science: Principles and Applications, 2nd edn. Springer, Fairfax (2017)
122. Crooks, A., Heppenstall, A., Malleson, N.: Agent-based modeling. In: Huang, B. (ed.) Comprehensive Geographic Information Systems. Elsevier, pp. 219–243 (2018)
123. Wilensky, U., Rand, W.: An Introduction to Agent-Based Modeling, 2nd edn. MIT Press, London (2015)
124. Railsback, S.F., Grimm, V.: Agent-Based and Individual-Based Modeling: A Practical Introduction. Princeton University Press, New Jersey (2012)
125. Heppenstall, A., Crooks, A.: Agent-Based Modeling in Geographical Systems, vol. 9 (2016)
126. Tobler, W.R.: A computer movie simulation urban growth in detroit region. Econ. Geogr. **46**, 234–240 (1970). https://doi.org/10.1126/science.11.277.620
127. Sullivan, D.O., Unwin, D.J.: Geographic Information Analysis, 2nd edn. Wiley, New Jersey (2010)
128. Gardner, M.: Mathematical games: the fantastic combinations of John Conway's new solitaire game "life". Sci. Am. 120–123 (1970)
129. Tobler, W.R.: Cellular geography. Philos. Geogr. 379–386 (1979). https://doi.org/10.1007/978-94-009-9394-5
130. Clarke, K.: Land transition modeling with deltatrons. Santa Barbara (2008)
131. Laila Ab Ghani, N., Abidin, S.Z.Z., Zamani, M.: Generating transition rules of cellular automata for urban growth prediction. Int. J. Geol. **5** (2011)
132. Schiller, F.: Urban transitions: scaling complex cities down to human size. J. Clean Prod. 1–10 (2015). https://doi.org/10.1016/j.jclepro.2015.08.030
133. Benenson, I., Torrens, P.: Geosimulation: automata-based modeling of urban phenomena. Wiley (2004)
134. Torrens, P.: How cellular models of urban systems work, p. 68 (2000)
135. Norte, N., Pais, A.: Cellular automata and urban studies: a literature survey. ACE Archit. City Environ. **1** (2007)
136. Aburas, M.M., Ho, Y.M., Ramli, M.F., Ash'aari, Z.H.: The simulation and prediction of spatio-temporal urban growth trends using cellular automata models: a review. Int. J. Appl. Earth Obs. Geoinf. **52**, 380–389 (2016). https://doi.org/10.1016/j.jag.2016.07.007
137. Santé, I., García, A.M., Miranda, D., Crecente, R.: Cellular automata models for the simulation of real-world urban processes: a review and analysis. Landsc. Urban Plan **96**, 108–122 (2010). https://doi.org/10.1016/j.landurbplan.2010.03.001

138. Batty, M.: Cellular automata and urban form: a primer. J. Am. Plan. Assoc. (1997)
139. Arsanjani, J.J., Helbich, M., Kainz, W., Boloorani, A.D.: Integration of logistic regression, Markov chain and cellular automata models to simulate urban expansion. Int. J. Appl. Earth Obs. Geoinf. **21**, 265–275 (2011). https://doi.org/10.1016/j.jag.2011.12.014
140. van Vliet, J., White, R., Dragicevic, S.: Modeling urban growth using a variable grid cellular automaton. Comput. Environ. Urban Syst. **33**, 35–43 (2009). https://doi.org/10.1016/j.compenvurbsys.2008.06.006
141. Haase, D., Haase, A., Kabisch, N., et al.: Actors and factors in land-use simulation: the challenge of urban shrinkage. Environ. Model Softw. **35**, 92–103 (2012). https://doi.org/10.1016/j.envsoft.2012.02.012
142. Basse, R.M.: A constrained cellular automata model to simulate the potential effects of high-speed train stations on land-use dynamics in trans-border regions. J. Transp. Geogr. **32**, 23–37 (2013). https://doi.org/10.1016/j.jtrangeo.2013.08.005
143. Gong, W., Yuan, L., Fan, W., Stott, P.: Analysis and simulation of land use spatial pattern in Harbin prefecture based on trajectories and cellular automata—Markov modelling. Int. J. Appl. Earth Obs. Geoinf. **34**, 207–216 (2015). https://doi.org/10.1016/j.jag.2014.07.005
144. Loibl, W., Toetzer, T.: Modeling growth and densification processes in suburban regions—simulation of landscape transition with spatial agents. Environ. Model Softw. **18**, 553–563 (2003). https://doi.org/10.1016/S1364-8152(03)00030-6
145. Sohl, T.L., Claggett, P.R.: Clarity versus complexity: land-use modeling as a practical tool for decision-makers. J. Environ. Manage. **129**, 235–243 (2013). https://doi.org/10.1016/j.jenvman.2013.07.027
146. Abar, S., Theodoropoulos, G.K., Lemarinier, P., O'Hare, G.M.P.: Agent based modelling and simulation tools: a review of the state-of-art software. Comput. Sci. Rev. **24**, 13–33 (2017). https://doi.org/10.1016/j.cosrev.2017.03.001
147. Epstein, J.M.: Agent-based computational models and generative social science. Complexity **4**, 41–60 (1999). https://doi.org/10.1002/(SICI)1099-0526(199905/06)4:5%3c41:AID-CPLX9%3e3.3.CO;2-6
148. Wilensky, U.: NetLogo flocking model. Center for Connected Learning and Computer-Based Modeling, Northwestern University, Evanston, IL (1998). http://ccl.northwestern.edu/netlogo/models/Flocking
149. Box, G.E.P.: Robustness in the strategy of scientific model building. In: Launer, R.L., Wilkinson, G.N. (eds.) Robustness in Statistics, pp. 201–236 . Academic Press, INC. (1979)
150. Grimm, V., Berger, U., Bastiansen, F., et al.: A standard protocol for describing individual-based and agent-based models. Ecol. Model. **198**, 115–126 (2006). https://doi.org/10.1016/j.ecolmodel.2006.04.023
151. Grimm, V., Berger, U., DeAngelis, D., et al.: The ODD protocol: a review and first update. Ecol. Model. **221**, 2760–2768 (2010). https://doi.org/10.1016/j.ecolmodel.2010.08.019
152. Sayama, H.: Introduction to the Modeling and Analysis of Complex Systems, 1st edn. Open SUNY Textbooks, Geneseo (2015)
153. Caparrini, F.S., García Vázquez, J.C.: NetLogo Una herramienta de modelado. CulturePlex (2017)
154. Vidal, J.: Fundamentals of Multiagent Systems with Netlogo examples (2010)
155. Lee, J.S., Filatova, T., Ligmann-Zielinska, A., et al.: The complexities of agent-based modeling output analysis. Jasss **18**, 1–26 (2015). https://doi.org/10.18564/jasss.2897
156. Lima, M.: Visual Complexity: Mapping Patterns of Information. Princeton Architectural Press, New York (2011)
157. Monks, T., Currie, C.S.M., Onggo, B.S., et al.: Strengthening the reporting of empirical simulation studies: introducing the STRESS guidelines. J. Simul. **7778**, 1–13 (2018). https://doi.org/10.1080/17477778.2018.1442155
158. Müller, B., Bohn, F., Dreßler, G., et al.: Describing human decisions in agent-based models—ODD+D, an extension of the ODD protocol. Environ. Model. Softw. (2013). https://doi.org/10.1016/j.envsoft.2013.06.003

Adaptive Security Based on MAPE-K: A Survey

Evangelina Lara, Leocundo Aguilar, Mauricio A. Sanchez and Jesús A. García

Abstract As systems evolve into interconnected heterogeneous components, their security threats increase in number and complexity, and static security measures are not capable of confronting all of them. A strategy to address this issue is the use of autonomic software, which adapts the security mechanisms at runtime according to the environmental changes that impact on the required security level. An approach to achieve autonomic computing is by using the MAPE-K reference model developed by IBM, which consists of a feedback loop composed of the functions: Monitor, Analyze, Plan, and Execute. In this manuscript, adaptive security models based on MAPE-K are surveyed, their characteristics are described, and a comparison of their domains, structures, and adaptive objectives is presented.

1 Introduction

Information systems have evolved into interconnected heterogeneous components, which makes it difficult for architects to anticipate their interactions at design-time. A solution is autonomic computing, which consists of systems that can manage themselves at runtime following high-level objectives from the administrators. The idea and its name originates from biological systems, which are composed of sev-

E. Lara (✉) · L. Aguilar · M. Sanchez · J. A. García
Facultad de Ciencias Químicas e Ingeniería, Universidad Autónoma de Baja California, 22390 Tijuana, BC, Mexico
e-mail: evangelina.lara@uabc.edu.mx

L. Aguilar
e-mail: laguilar@uabc.edu.mx

M. Sanchez
e-mail: mauricio.sanchez@uabc.edu.mx

J. A. García
e-mail: garcia.jesus@uabc.edu.mx

M. A. Sanchez et al. (eds.), *Applied Decision-Making*, Studies in Systems, Decision and Control 209, https://doi.org/10.1007/978-3-030-17985-4_7

eral self-governing components that interact with each other, such as the autonomic nervous system which governs the heart rate, freeing the conscious brain from this low-level and vital function. The primary purpose of autonomic systems is the capability of self-management, that is, to be able to adapt their operation on changing environments, components, and workloads, and also, in the presence of hardware and software failures; thereby freeing administrators from their configuration, integration, and maintenance [1].

An approach to achieve autonomic computing is by using the MAPE-K reference model [2], developed by IBM in 2005. Its name originate from the first letter of the phases that compose it: Monitor, Analyze, Plan, and Execute, and the K corresponds to the Knowledge shared between the four phases. Monitor phase is responsible for collecting managed resource's details, data correlation, and determination of symptoms that need to be addressed. Analyze phase performs an analysis and reasoning on the data provided by the monitor stage; if it determines that a change is needed, it sends a petition to the Plan stage. Once receiving the request, Plan phase structures the required actions to achieve goals and objectives. It creates or selects procedures to declare alterations that have to be made on the managed resource. Then, Execution phase changes the behavior of the managed resource according to the recommended actions of the planning phase. Knowledge consists of standard data shared between the phases. It is created by Monitor phase, and Execute can update it.

MAPE-K is a generic model for autonomic computing; it can be applied to diverse system types to achieve the capability of adjusting their behavior at runtime according to environmental changes. It is often used as a feedback loop [3] for the dynamic adaptation of systems such as cyber-physical systems [4, 5], opportunistic sensing [6], monitoring and control systems [7], safe and reliable embedded systems [8], security [9–18], etc. This manuscript centers on one of them, which is adaptive security systems.

As systems evolve, their security threats increase in number and complexity, and static security measures are not capable of confronting all of them. A strategy to address this issue is the use of adaptive software which adapts the security mechanisms at runtime according to the environmental changes that impact on the required security level. Therefore, a system can tighten their security policies and mechanisms in the presence of attacks, or modify them in case of legal or organizational changes. Several approaches for adaptive security have been proposed, which have different architectures and procedures.

The rest of this chapter is composed as follows; the next section contains the background. Section 3 presents the survey of the adaptive security systems. A discussion is in Sect. 4. Finally, conclusions are given in Sect. 5.

2 Background

2.1 Autonomic Computing

The base of autonomic computing is self-management, which comprises four aspects:

- Self-configuration: Installing and integrating large systems is a complex and error-prone task. Autonomic components will incorporate themselves into systems transparently, according to high-level policies which indicate what have to be accomplished, not how to do it.
- Self-optimization: Autonomic systems will monitor and experiment with their parameters with the purpose of improving their efficiency, performance, or cost.
- Self-healing: Autonomic systems will identify, diagnose, and correct software and hardware failures.
- Self-protection: Autonomic systems will forestall, prevent, and mitigate system-wide failures, and defend themselves against malicious intrusions, or cascading breakdowns.

Autonomic systems will be composed of autonomic elements, which manage their behavior according to policies of administrators or other system components, and provide services to humans and other elements. The development of autonomic systems has several challenges, as can be seen in the life cycle of a single autonomic element, which comprises five phases:

- Design, test, and verification: Developers will require tools for representing the high-level goals and limitations of the autonomic element, and translate them to low-level actions. Testing autonomic elements of large-scale systems will be challenging because it will be difficult to anticipate or coordinate the interactions of all the autonomic elements with which they have a dependency, or build test systems that accurately represent the complexity of those systems.
- Installation and configuration: The installation process will begin with the registration of the autonomic element in a directory which contains the offered services of each of the components in the autonomic system. The element can also use the directory to discover suppliers that can help it to complete its installation and configuration, according to the policies of the system of which it will be part.
- Monitoring and problem determination: Monitoring is an essential part of autonomic computing. Elements will regularly monitor themselves, their suppliers, and customers to ensure that the agreed service level is being received. Monitoring support problem anticipation, detection, and recovery, but it will be a challenge to perform those actions without having a high impact on the system. When detecting a failure, it will also be challenging to recover quickly and gracefully without affecting the rest of the system elements.
- Upgrading: Autonomic elements might subscribe to services which inform them of available upgrades, and they will determine when to apply them. The autonomic system can also decide the replacement of some of its elements for new ones as part

of the upgrading; the old elements will be eliminated after the system determines that the new ones are performing properly.

- Uninstall: The life cycle of an autonomic element concludes when it is removed or replaced from the system.

The interaction between autonomic elements also has challenges, which can be seen in its life cycle:

- Specification: Autonomic elements provide a set of services, and consume a set of services from suppliers. This information has to be communicated in standard formats to allow other autonomic elements to be able to interpret and reason it.
- Location: Autonomic elements must be able to find the services that they require, advanced reasoning might be needed to determine which supplier can provide the exact services with reliability.
- Negotiation: After an autonomic element finds a service provider which can fulfill its necessities, it has to negotiate with the supplier to obtain it. Both parties have to reach an agreement about service level, price, and priority. Negotiation strategies and languages for communicating service agreements are needed.
- Provision: Once provider and consumer of a service reach an agreement, they have to provision their resources, which can consist of saving a record of the element or service name it can access in the future, or it may be necessary to look for other elements to act as subcontractors for part of the agreed service.
- Operation: The elements operate according to the negotiated agreement. The autonomic managers of the service provider and service requester monitor the service level to ensure that the agreement is being respected.
- Termination: When the agreement has finished, the parties accord to end it. The elements may save information relative to the service relationship, or store it in a reputation database.

Another difficulty on the development of autonomic systems corresponds to the methods to ensure that the high-level policies are correctly stipulated. Although these systems reduce human errors, if the goals are not set correctly, their consequences are magnified. Therefore, there have to be methods to ensure that policies are correct and that the system performs reasonably even if they are not [1].

The challenges on the specification of adaptive software are [19]:

- Notation formalization: The absence of formalization in the notation used for requirement specification causes issues such as allowing the definition of requirements in more than one way, introducing ambiguity.
- Traceability between requirement elicitation and specification: Forward and backward traceability facilitates stakeholders the review and update of the elicited requirements and their specification. Requirement elicitation usually obtains requirements in natural language; thus, achieving traceability from these non-formal documents to specification models is a complex task.
- Restriction definition: The restrictions when defining relations and controlling transitions from one state to another involve several operations and variables. Support for rich constrain definition is needed to prevent unwanted transitions.

- Formalization of verification operations: An specific syntax for verification operations, and support for the definition of new ones is required.
- In-depth verification: Verification models should involve distinct concern levels, including subsystems and subsystems within other subsystem.
- Transition verification: State graphs and transitions should be validated to identify non-reachable or blocking configurations, and non-stable or undesired adaptations.
- Optimization models and verification: Optimization models should include ones that satisfy attributes such as cost, risk, and performance. Their evaluation should be able to combine different models for a more realistic verification, solving and balancing contradictory results.
- Resilience verification: Considering that not all the system features may be functional all the time, resilience verification needs to identify features prone to be non-functional and the environments where those should be used, and evaluate if alternatives for them exist.
- Optimal accuracy verification: A subset of system goals must be satisfied in all the environments for a minimum operative configuration. The optimal accuracy operation validates that in every environment there exists at least one configuration that comply with the collection of required goals, with some tolerance level.
- Overhead verification: Adaptation should have a small resource overhead to be able to complete it on time. The challenge is how to verify, from the specification, that the lowest cost is achieved in all of the expected adaptations.

Furthermore, autonomic systems will have the security and privacy problems that traditional computing systems confront, but their consequences could be more severe because if attackers manipulate high-level policies, they could obtain more influence on the system's components than is possible in traditional systems [1].

A methodology for the achievement of autonomic behavior is MAPE-K reference model, briefly presented in Introduction, which consists of a loop structure that supervises a resource, detects environmental changes that affect it, and decides the best way to address or adapt to those. Next subsection describes this model.

2.2 MAPE-K Reference Model

MAPE-K reference model is composed of four functions:

- Monitor: It collects information about the managed resource and its environment, such as configuration settings, status, metrics, etc. The data can be static or dynamic. It uses sensors, which can be of hardware or software, to obtain operation details and environmental changes. It aggregates and correlates the information to find symptoms and the combination of events that generated them. It sends to Analyze function the symptoms, for their analysis and determination if changes on the managed resource are needed.
- Analyze: It contains mechanisms to analyse and reason the information sent by Monitor, and decide if changes in the behavior of the managed resource are

required. A change may be necessary if a policy is not being met. This function can utilize prediction mechanisms, such as queuing models, to help it anticipate future behavior. If it decides that a change in the managed resource is needed, it sends to the Plan function a request, which contains the modifications that Analyze considers necessary.
- Plan: It generates a change plan by creating or selecting procedures for the accomplishment of the modifications on the managed resource, and sends it to Execute function.
- Execute: It schedules and performs a series of actions to modify the managed resource, according to the received change-plan of Plan function.

The functions have a shared Knowledge, which consists of policies, symptoms, and change requests and plans. It can be implemented as a registry, dictionary, or database, and it can also be shared between distinct autonomic managers. The knowledge can be obtained in one of this means:

1. It is transferred to the autonomic manager. The autonomic managers obtain the policies in this manner.
2. It is acquired from external knowledge sources. They might contain symptoms and events that have happened in other components or systems and can be used by the autonomic manager.
3. It is created by the autonomic manager and stored within it. Monitor function creates it by logging the sensed data and the notifications received from the managed resource. Execute function updates it by indicating the actions that were performed on the managed resource as a result of Analyze and Plan phases, and how those operations affected it. The information has to be located in a knowledge source to be able to share the knowledge with other autonomic managers.

The knowledge that may be present in an autonomic system can be categorized into three types:

- Solution Topology: It consists of the knowledge related to the construction and configuration of the system's components. It can be used by Plan function when planning the installation and configuration of an element.
- Policy: It consist of high-level goals and restrictions. It is consulted by Analyze function to decide if changes in the managed resource are required. It should be defined in a standard format to allow it to be shared between autonomic managers, to accomplish the complete system to be controlled by a common set of policies.
- Problem Determination: It consists of data collected by Monitor function, symptoms, decision trees, and learned knowledge about the system response to actions performed to correct problems.

The autonomic manager uses sensors to monitor the managed resource, and effectors for its modification according to the change plan developed by Plan function. They can be of software or hardware, and present their functions as manageability interfaces to the autonomic managers. A sensor contains one or both of these interaction styles:

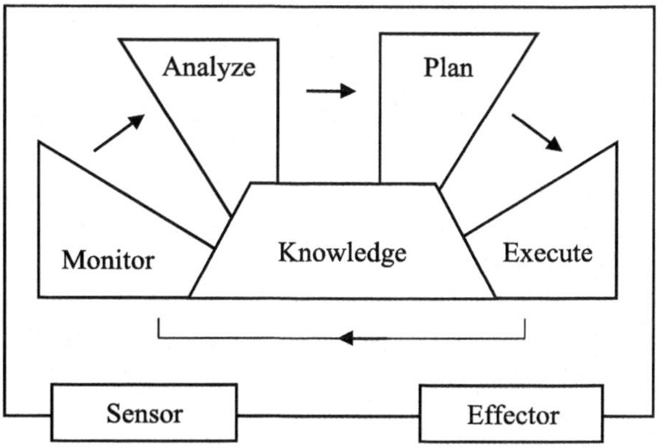

Fig. 1 The MAPE-K loop structure for autonomic computing

- Properties that expose the status of the managed resource, they are obtained through get operations.
- Management events that execute when the managed resource experiences significant changes, e.g., sending the event information to autonomic managers.

An effector contains one or both of these interaction styles:

- Set operations that change the state of the managed resource.
- Operations that are used by the managed resource to make requests to its manager to get services from other entities in the system.

In Fig. 1 the structure of the model is presented, it contains the functions, elements, and feedback loop that compose a resource which have autonomic capabilities [2].

2.3 Adaptive Security

Dynamic and heterogeneous systems are challenging to provide security because static security mechanisms could not be adequate for all of the variable scenarios, and it is not always possible to anticipate all the contexts and conditions in which the systems will be running. Furthermore, as systems evolve, their security threats increase in number and complexity. Adaptive security can provide a solution to systems where static security measures are insufficient [11, 20] because they autonomously and at runtime adapt their security mechanisms and their configurations according to the current state and needs of the system and its environment.

The challenges to the implementation of adaptive security consist of [13]:

1. Because dynamic systems are composed of heterogeneous elements, the security mechanisms developed for one element have to be able to be adapted to the characteristics of all of the other components.

2. The conditions on the system's environment can change, and the security policies have to be adapted to them. The adaptation requires a dynamic negotiation of new policies by the different elements that compose the system.
3. The context has to be monitored for changes to be able to react to them, because those can have an impact on security.
4. Systems have to have self-protection mechanisms; thus, if the context or security requirements change at runtime, the systems conserve an appropriate security level and functioning.
5. Dynamic systems consist of a large number of elements; thus, the security mechanisms have to be effective and efficient when running on them.

To achieve adaptive security through MAPE-K, the four functions of the model may consist of [11]:

- Monitor: It examines the system and its environment looking for context and security changes.
- Analyze: Analyses the impact of the changes detected by Monitor on the security objectives and policies. To do it, not only the current security level has to be known, but also the required one. The required security level for the system at the moment can be specified at design-time, or it can be reasoned by Analyze function. Several aspects have to be modeled to accomplish the most appropriate approach, such as high-level goals, current context, threats, available resources, etc. If it determines that a change on the security measures is necessary, it sends a change request to Plan function.
- Plan: It creates and structures an adaptation plan according to the modifications indicated by Analyze, which includes the selected security mechanisms.
- Execute: It effectuates the adaptation plan.

To achieve the security objectives of a system, may be necessary to comply, at least, with these three services [21]:

- Authentication: It is the process through which an entity verifies the identity of a second party on the protocol. It also serves the security objective of non-repudiation, which consists of preventing a user from denying its performed activities and acquired responsibilities. Adaptive authentication consists of selecting the most appropriate authentication method according to the current system state, user type, and the user associated trust-level.
- Authorization: It is the management of access to system services and resources. It includes the assignment of access rights to users. It supports the security objectives of confidentiality and integrity, by ensuring that only authorized users access and modify their authorized resources. Adaptive authorization consists of selecting the most appropriated access control policies according to the current system state, the user, and the resource.
- Tolerance: It is the persistence of the system even in the presence of system failures or intrusions. It consists of containing the fault impacts or hiding faults by

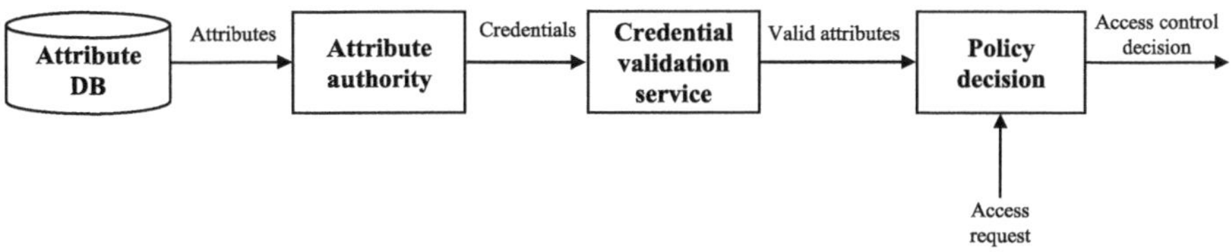

Fig. 2 The RBAC/ABAC model on which the adaptive authorization framework is based

eliminating or substituting faulty components. It supports the security objective of availability, which is the ability to keep providing services with an acceptable service level even when faults occur. It is inherently an adaptive process.

3 Survey of Adaptive Security Systems

A survey of ten adaptive security systems is presented next.

3.1 Adaptive Authorization Framework, Bailey et al.

In [10] an Adaptive Authorization Framework for RBAC/ABAC [22, 23] systems is presented. Its purpose is monitoring the usage of the authorization infrastructure, reasoning about the users' behavior concerning the authorization requests, and adapting the infrastructure appropriately. The framework is reactive, i.e., if an unusual behavior is detected, the system autonomously decides if the authorization infrastructure has to be changed.

In Fig. 2, the RBAC/ABAC model on which the framework is based is shown. As can be seen, user attributes determine what credentials will be granted, which restrict the user access permissions. Through changing the attributes, the access permissions can be reduced or incremented. Also, credentials can be revoked when the holder is in an access session, consequently terminating it.

The infrastructure uses a policy model where each authorization policy has a meta-policy, which contains behavioral rules that represent usage thresholds, that can allow the detection of abnormal permission usage. The behavioral rules have associated cost functions which indicate the impact to the organization if a rule is broken. The infrastructure also depends on usage statistics, which contain historical usages such as access requests and access control decisions. The statistics allow the system to make conclusions about roles, attributes, subjects, and permissions.

In Fig. 3 the architecture of the adaptive framework is presented, which consists of the functions of MAPE-K model. The policy model and usage statistics correspond to the shared knowledge.

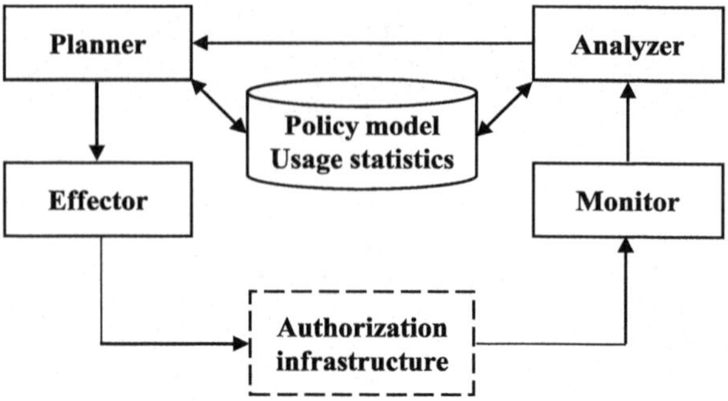

Fig. 3 The architecture of the adaptive authorization framework of Bailey et al

The MAPE-K functions are mapped on the framework in this manner:

- Monitor: It obtains from the authorization infrastructure the access requests and the authorization decisions, and sends them to the analyser component.
- Analyzer: It processes the monitored data and tries to identify abnormal behavior using the policy model and usage statistics. It updates those knowledge sources with the analysed data for future reference. When abnormal behavior is detected, it provides solutions to the planner component to prevent or accept it, which can consist of modifying authorization rules or adding/removing user attributes.
- Planner: It selects a cost-effective solution from the set of solutions sent by the analyser component, creates a plan for it, and sends it to the effector component. It updates the policy model according to the selected solution.
- Effector: It adapts the infrastructure according to the received plan, which can consist of modifying the user attributes and credentials, or the management of the authorization policies.

As can be seen in Fig. 3, only the monitor and effector components know the particular interfaces, protocols, and policy formats used by the authorization infrastructure.

The adaptation process of the authorization infrastructure comprises four activities: trigger adaptation, solution analysis, solution selector, and plan generation. The first two are performed by the analyser component, and the last two by the planner, as can be seen in Fig. 4.

- Trigger adaptation: It analyses the user's system usage according to the policy model. From his/her access requests can be identified its active time, the request rate, the number of accepted and denied requests, etc., this information is used to specify behavioral rules to permissions and attributes. If the user does not follow the behavioral rules related to the current permission, the solution analysis is started.
- Solution analysis: It interprets the user behavior according to the policy model. Solutions can consist of alternative rules that prevent or support the abnormal behavior.

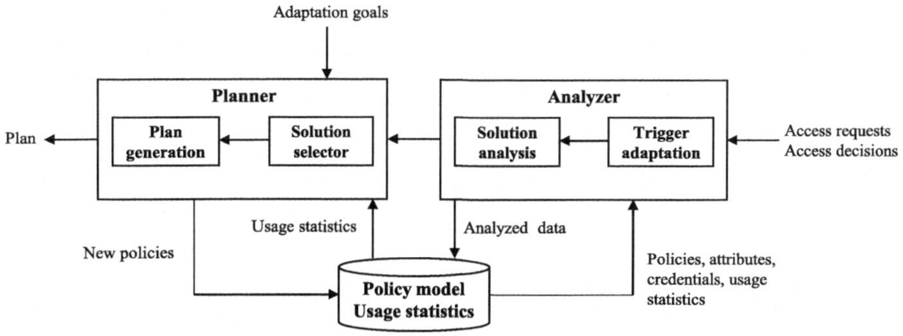

Fig. 4 The adaptation process of the adaptive authorization framework of Bailey et al

- Solution selector: It evaluates which solution best met the adaptive goals. Goals can be just one attribute, such as cost, then the selected solution may be the most cost-effective.
- Plan generation: It determines the required actions to perform the selected solution. The plan consists of step-by-step instructions to achieve the adaptation.

3.2 Adaptive Security in Smart Space, Evesti et al.

In [11] an architecture for adaptive security is presented, it is intended for smart spaces, which are inherent dynamic systems. It is composed of three elements:

- Adaptation loop: It comprises the four functions of MAPE-K model. It is responsible for the selection of appropriated security mechanisms and their parameters at runtime.
- Input knowledge: It consists of an Information Security Measuring Ontology (ISMO) [24], which is used by the adaptation loop to decide if something has to be changed. It supports the addition and modification of knowledge, and minimizes the required hard-coded information.
- Security-control model: It enforces the dynamic policies of access control and authorization at runtime.

The MAPE-K functions are mapped on the adaptation loop in this manner:

- Monitor: It collects information of the smart space application and its environment. The monitored attributes can be of low-level, such as network technology, cryptographic keys, etc., or high-level, such as the user state. It uses Monitoring probes, which are concrete code sections, to observe a particular attribute. The knowledge of the ISMO is used to decide which probes to use at a certain moment, e.g., if a situation does not need integrity in the communication, it is not necessary to use integrity-related probes.

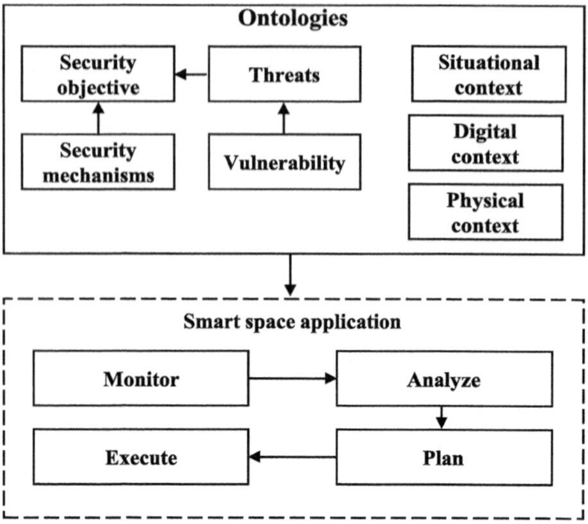

Fig. 5 Ontologies and adaptation loop of the adaptive security model for smart spaces proposed by Evesti et al

- Analyze: Monitor function detects the changes in the smart space, whereas Analyze function determines their consequences in the required security goals of the current situation. It compares the current security levels of high-level security goals against the required ones, if the current levels do not comply, Plan function is initiated. It uses Indicators, which obtain and present the security level of a specific security objective, e.g., authentication, with which it is linked.
- Plan: It decides how to adapt the smart space application when the security levels are not met. It uses the results of Analyze function, such as the unsatisfied security objective, the required and current security level, the security mechanism used, etc., and uses the ISMO to build an adaptation plan. Three possibilities exist for its creation:
  - The configuration alternatives are pre-defined.
  - The configuration alternatives are created using the knowledge of the ISMO. The smart space application searches for alternative security mechanism on it.
  - The smart space application inquires the user about how to proceed.
- Execute: It performs the adaptation plan. It can have an impact on different layers of the smart space, e.g., when a communication protocol is adapted, the smart space application may need to create a new connection with previously connected applications, using new mechanisms or parameters.

Figure 5 presents the adaption loop and the elements that compose the ontologies, used by the adaptation loop to determine if changes are required.

3.3 Adaptive Access Control Model, Ma and Wang

In [9] an adaptive access control model for cloud computing is proposed, which consists of a feedback loop. A characteristic that distinguishes this model is that it also proposes an algorithm to generate the knowledge base used by Analyze function to determine if a change is required.

The knowledge base stores and manages prior knowledge to support access control decisions. It computes and stores the relation degree between access control attributes. Those relations repair, improve, and gradually construct an intelligent knowledge base. The construction consists of taking historical access control decisions and applying data mining methods to obtain the relationship between the involved access control attributes. Then, when an access control request is received, the relationship knowledge is used to make an access control decision. This process is repeated as a feedback loop, as showed in Fig. 6.

A feedback control loop is used to achieve an adaptive access control system, which monitors the system to obtain its state and then compares it with the expected one to decide if something has to be changed. The feedback loop uses the knowledge base, and its components consist of the functions of MAPE-K model, mapped in this manner:

- Monitor: It observes the historical records and the current access requests, attributes, and behavior.
- Analyze: It examines the access behavior to decide if it is necessary to update the knowledge base, and selects historical records for the knowledge base updating.
- Plan: It calculates the relationship degree of the historical records sent by Analyze function, and updates the knowledge base to support the access control decision-making process.
- Execute: It uses the knowledge base to make appropriated decisions on the received access requests.

The relationship between the components of the adaptive access control model is shown in Fig. 7.

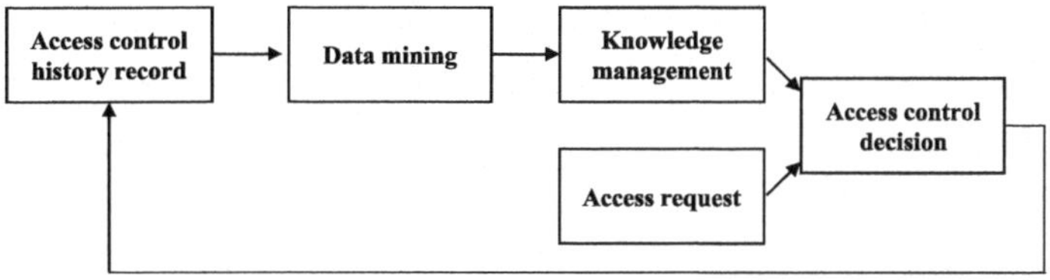

Fig. 6 The construction process of an intelligent knowledge base, proposed by Ma and Wang

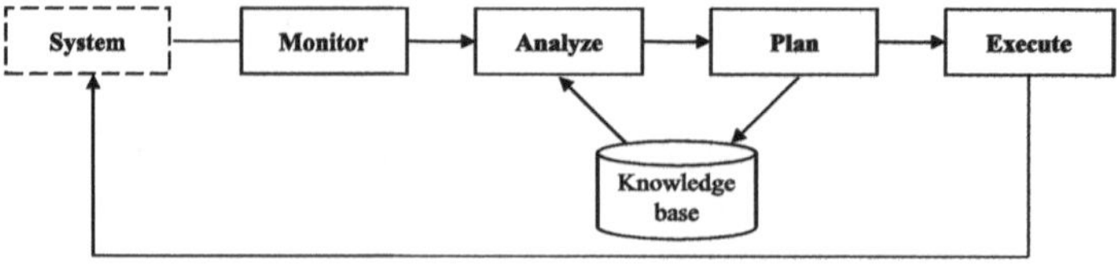

Fig. 7 The feedback loop of the adaptive access control model proposed by Ma and Wang

3.4 Dynamic Reconfiguration of Security Policies, Pinto et al.

In [12] is presented a model for self-protection in Wireless Sensor Networks (WSNs). It is based on Dynamic Software Product Lines (DSPL) [25, 26], which generate software products able to adapt to requirements that change at runtime by modeling their elements as dynamic variation points. The adaption of security policies at runtime implies their modeling and negotiation; thus, the INTER-TRUST framework [27, 28], and FamiWare middleware [29] are used by the authors. INTER-TRUST is a framework for the dynamic negotiation and enforcement of security policies at runtime. However, it does not get involved with the variability of the WSN structure. FamiWare is a DSPL for the automatic configuration of middleware instances; it provides a solution for the modeling and reasoning of the WSN heterogeneity. These two software engineering technologies are used to hide the WSN details from application developers and to perform the autonomic behavior when security requirements change.

The relation between MAPE-K model and the security features of the integration of INTER-TRUST and FamiWare include these:

- Monitoring: It examines problems related to security. It is related to Monitor function of MAPE-K.
- Context awareness: It decides if a security policy has to be negotiated and the required changes in the system security in case it has to. It is related to Analyze and Plan functions of MAPE-K.
- Reconfiguration: It reconfigures the WSN nodes by executing the reconfiguration plan generated by Context awareness. It instantiates new security policies in the appropriated sensor nodes. It is related to Execute function of MAPE-K.

In Fig. 8 the components of the integration of INTER-TRUST and FamiWare related to adaptive security are shown.

3.5 Adaptive Security Solution, Amoud and Roudies

In [13] an architecture for adaptive security is presented, which maps the components of MAPE-K reference model to five units that enable dynamic systems to select and configure security mechanisms at runtime.

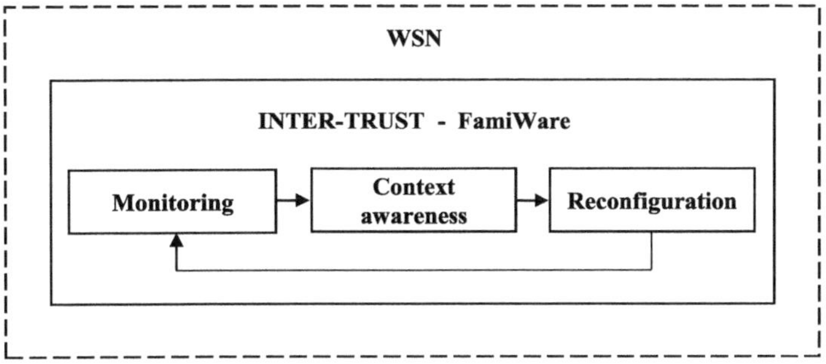

Fig. 8 The components of the integration of INTER-TRUST and FamiWare for adaptive security in WSNs, proposed by Pinto et al

- Context unit: It corresponds to Monitor function. Its purpose is the perception of the current context of the system and its environment, for supporting Analyze-Decision unit to make informed and smarter decisions.
- Analyze-Decision unit: It corresponds to Analyze and Plan functions. Its purpose is the determination of the future system state. It determines if security measures should be changed, and structures the required actions to achieve high-level goals, by creating a policy that conforms with the adaptation goal.
- Adaptation unit: It corresponds to Execute function. It is responsible for executing the security adaptation at runtime. It intercepts the input data and maps the policy rules to it by specifying to the Security unit which security functions have to be applied to it.
- Security unit: It applies to the input data the security functions indicated by Adaptation unit and returns the resulted data.
- Knowledge unit: It corresponds to the shared Knowledge. Consists of data such as the current state of the system and environment, services' state which contain a system model of the interrelationship between the environment situation and security, and policies which conform with the adaptation goal.

Figure 9 presents the architecture for adaptive security, with the five units that compose it, their internal elements, and the relationship between the components.

3.6 Adaptive Security in Mobile Devices, Amoud and Roudies

In [14] an adaptive security model for mobile devices is presented. It uses the MAPE-K reference model to negotiate and deploy security policies, and DSPL for context monitoring to react to its changes and to configure the security level on the mobile device's applications. Also, it uses the Common Variability Language (CVL) [30] to validate the proposed model.

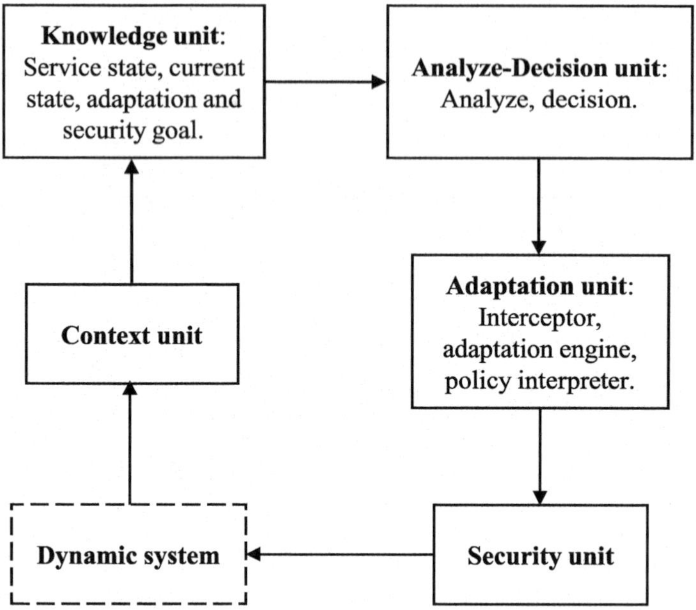

Fig. 9 The architecture for adaptive security on dynamic systems, proposed by Amoud and Roudies

The components of the adaptive security model are shown in Fig. 10, which correspond to the elements of MAPE-K model mapped in this manner:

- Context unit: It perceives the system and environment context, enabling the Analyze-Decision unit to make informed decisions.
- Analyze-Decision unit: It is composed of two elements: Analyze and Decision. Analyze structures the required actions to achieve goals. Decision determines if security needs to be adapted according to the adaptation goal, and how to do it. The Analyze-Decision unit creates a policy which conforms with the adaptation goal.
- Adaptation unit: It is composed of three elements: Interceptor, Policy Interpreter, and Adaptation Security. The first one receives the source data directed to the Security unit. Policy Interpreter translates the updated policies generating Security function and Data filter pairs that are input to the Adaptation engine, which applies those conditions to the data from the Interceptor. The filtered data are then sent to the Security unit.
- Security unit: It performs the Security functions on the input data and returns the resulted data.
- Knowledge unit: It contains data of the current state and the relationship between security and the environment.

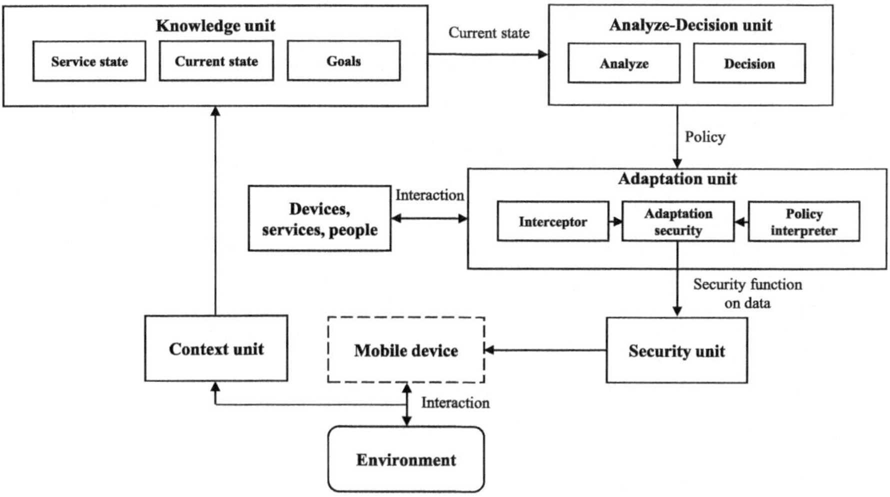

Fig. 10 Components of the adaptive security model proposed by Amoud and Roudies

DSPL is used by the proposal to solve modeling problems associated with the variability and heterogeneity of mobile devices, it allows the creation of configurable products, enabling devices to configure and reconfigure middleware instances at runtime. Also, CVL is used for the specification and resolving of the system variability. It allows modeling the base and variability models separately, but connected and managed using the same tool. Variability models guide system components on their reorganization to move from a system configuration to another. In Fig. 11 the modules of the variability model are presented, which consist of the following:

- Context awareness: It is composed of the security analysis and product generation features. The first one determines if security policies should be negotiated and the required changes on the system if they have to. It includes the policy interpreter and policy engine. Product generation indicates that a new configuration has to be created, considering the dynamic information of the policy interpreter component.
- Reconfiguration: It reconfigures the system security by instantiating the new security policies in the required mobile devices of the network.
- Negotiation: It is responsible for the dynamic negotiation of security policies.

3.7 Adaptive Security in Smart Camera Networks, Amoud and Roudies

In [15] an adaptive security model for Smart Camera Networks (SCN) is presented. Its architecture is equivalent to the one proposed in [14], which was described in the previous subsection. In the manuscript, the authors propose its usage to achieve

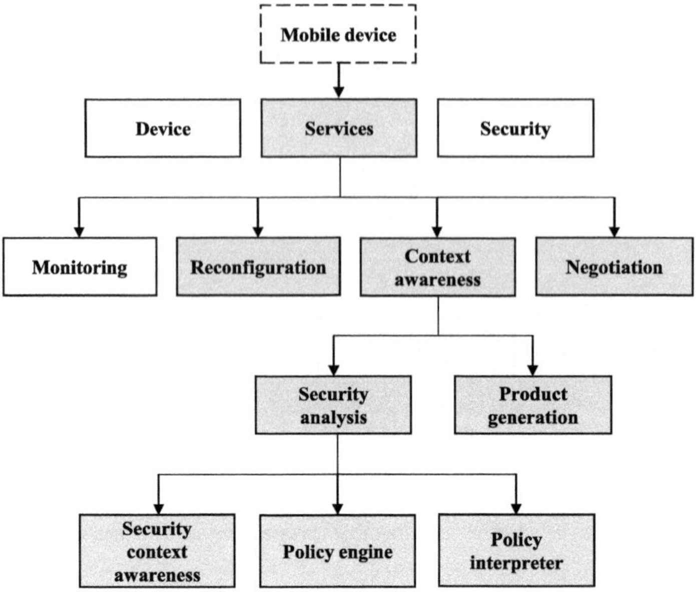

Fig. 11 Modules of the variability model proposed by Amoud and Roudies

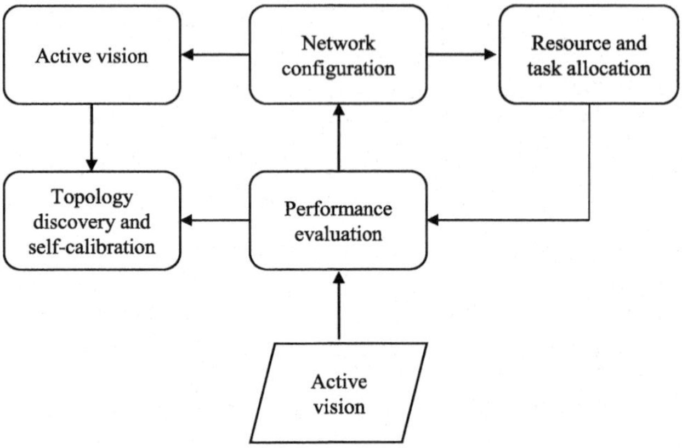

Fig. 12 Cycle of self-configuration in a SCN [31]

adaptive security in SCN, which consist of heterogeneous devices working on changing environments. SCN applications are dynamic and demand cameras to adapt to variable conditions, self-configuration allows them to collaborate to address unexpected environment or security conditions, such as changes in the communication bandwidth and intrusions [31]. Figure 12 presents the cycle of self-configuration in an SCN.

3.8 Adaptive Security in Intelligent Vehicles, Singh and Kim

In [16] an adaptive security mechanism for intelligent vehicles (IV) is presented. Some of the security challenges in IV environments are:

1. Steal of IDs.
2. Spread of wrong messages.
3. Message tampering.
4. DoS.
5. Verification of the legality of the vehicle.

The authors propose the following services to solve the challenges:

1. Liability identification.
2. Message authentication.
3. Message integrity.
4. Non-repudiation.
5. Message confidentiality.

The proposed model uses MAPE-K to achieve a secure communication between two IVs which need to exchange information. For example, in a context where an IV, denoted IV-A, receives a message from other IV, named IV-B, the MAPE-K functions act in this manner:

- Monitor: The IV-B identity is verified in this phase to know if it is a legal vehicle, i.e., it is registered with a Certificate center.
- Analyze: If the previous phase was successful, in Analyze phase the message is authenticated.
- Plan: If authentications are correct, IV-A creates a plan for a privacy-preserving and secure communication.
- Execute: IV-A communicates with IV-B according to the plan created in the previous phase.

Figure 13 presents the function blocks that compose the adaptive security model.

The model has two entities which make possible the secure communication, a self-diagnosis agent and a Certificate center. The former has predefined actions for the decision-making process for a secure connection. The latter is responsible for certificate issuing, communication monitoring, and to provide security-related information about IVs. Figure 14 shows each one of the responsibilities of these two entities.

3.9 Adaptive Access Control Model in Internet of Things (IoT), Gwak et al.

In [17] a role-based access control for public IoT environments is presented. The model is trust-aware, a characteristic that distinguishes it is that it can determine

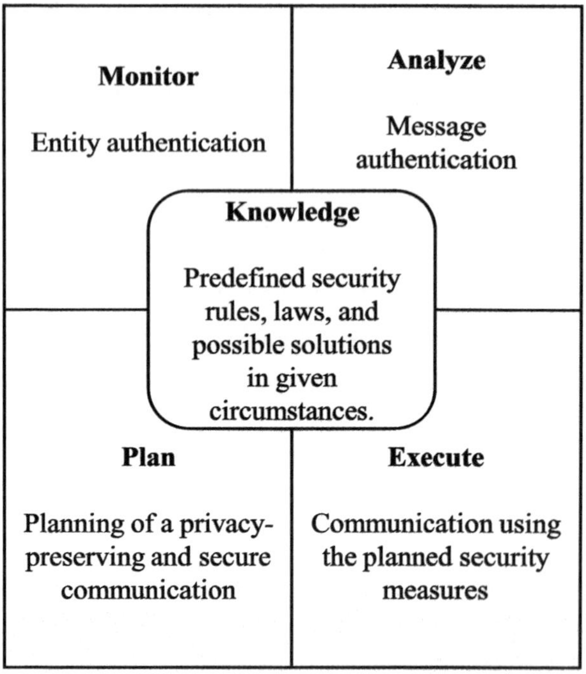

Fig. 13 Function blocks of the adaptive security model of Singh and Kim

an initial user's trust level and grant user's access rights without prior interaction of him/her with an smart object, which is convenient because in public networks new users arrive and leave continually. For the initial trust establishment, the model follows the rationale of I-sharing [32, 33], a psychology concept which says that people with similar characteristics and roles are inclined to respond similarly under given situations. After the initial estimation, the model can dynamically adapt the trust-level and revoke access rights of malicious users based on their interactions with the smart object.

The intended environment is a public IoT context, and the involved entities are a user, which is the Service Requester (SR), and an smart object, which is the Service Provider (SP). The sequence of steps that has to follow the SR to obtain a service from the SP is illustrated in Fig. 15, consisting of:

1. Two Factor Authentication with a Trusted Third Party (TTP): The SR authenticates with a TTP to obtain a list of available trusted SPs.
2. Role identification: After the SR chooses a particular SP, he/she is asked to provide his/her Social Networking Service (SNS) credentials for allowing the SR to access its SNS account. By accessing the account, the SR is able to verify the SP identity, and identify his/her role based on his/her information, such as education, interests, etc. The SP uses the role to get an I-sharing group for the SR, considering the users with which he/she has previously interacted.

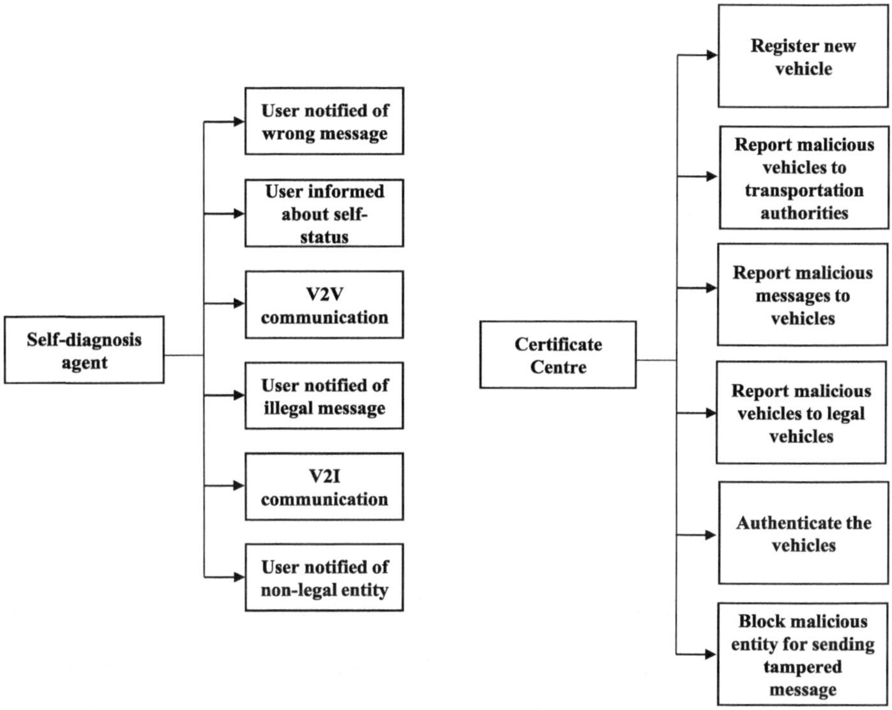

Fig. 14 Responsibilities of the self-diagnosis agent and Certificate center

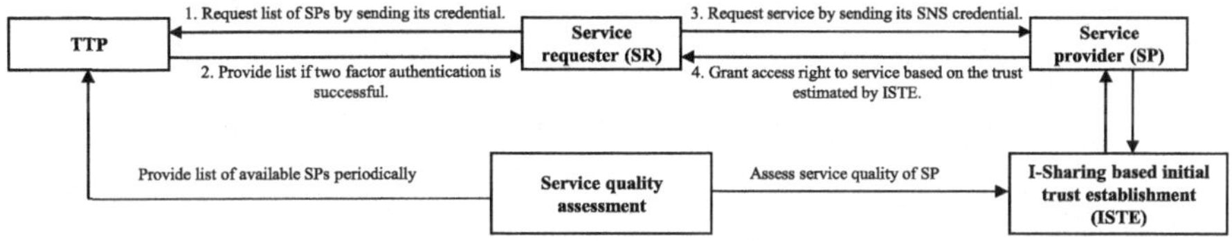

Fig. 15 Sequence of steps followed by a SR to obtain a service from a SP

3. Granting of access rights: Access rights are granted if the SR passes the previous authentication processes.

The model architecture is shown in Fig. 16, which consists of four managers: session, authorization, adaptation, and trust. The session and authorization managers are responsible for the authentication processes, allowing the SR to receive the lists of trusted SPs, and granting access rights to services of an SP. The adaptation and trust managers are responsible for ensuring the trustworthiness of the SR based on its interactions with the SP, and delivering the trust information to the authorization manager for the determination of the SR's access rights. The I-Sharing Group Manager (ISG) searches for a list of users who have interacted with the SP and identifies those who have similar characteristics with the SR. The ISG uses the trust values of

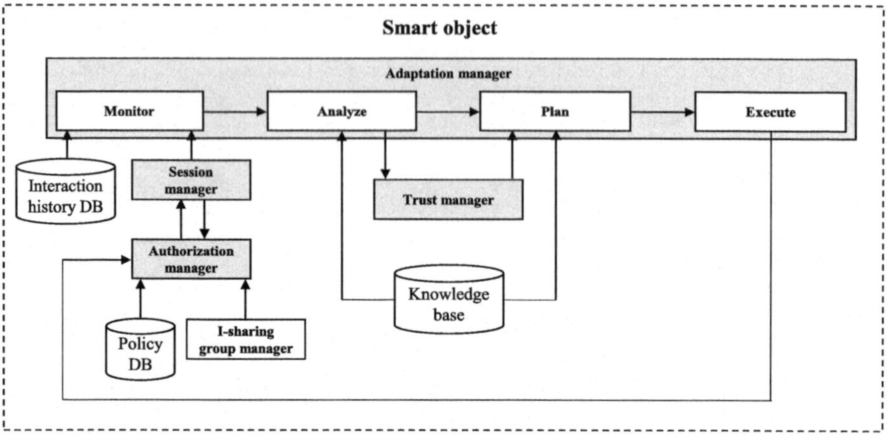

Fig. 16 Architecture of the adaptive access control model of Gwak et al

those users to establish the one of the SR, with which the SP does not have a previous interaction.

The adaptation manager uses the MAPE-K model for the dynamic adaptation of the user trust-level according to his/her interaction with the smart object, the four functions of the reference model are implemented in this manner:

- Monitor: It captures interaction information from the session manager, such as SR's ID, time and content of the interaction, and authorization decision.
- Analyze: It analyses the interaction information received from Monitor and determines if the SR's behavior is or is not malicious. The decision criteria is defined in a behavior policy.
- Plan: It examines the knowledge base for a policy that can be applied to the current situation, based on the SR's ID and his/her trust level, and generates a plan according to the policy.
- Execute: It modifies the authorization policy according to the plan generated by Plan phase.

3.10 Anomaly Detection in Cyber Physical Production Systems (CPPS), Settanni et al.

In [18] an anomaly detection mechanism to enable adaptive security in CPPS is presented. It intends to monitor an industrial system and learn its normal behavior, and generate alerts when abnormal activities are detected. The authors propose the use of this type of mechanism to support the phases Monitor and Analyze of MAPE-K model, where anomaly detection techniques monitor and analyses events on the CPPS and trigger alarms if unusual activities are identified, with the purpose of

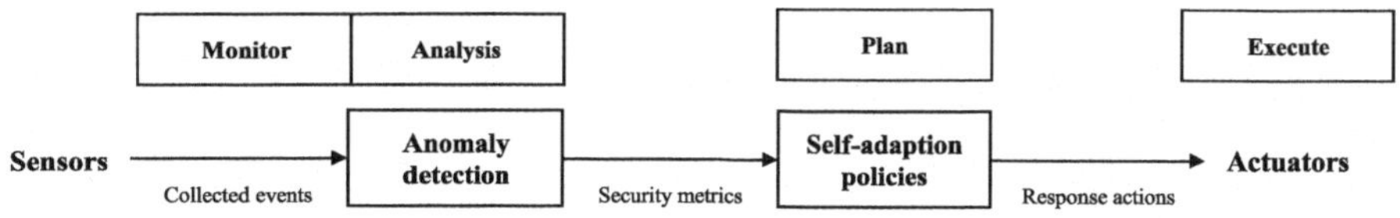

Fig. 17 Anomaly detection for adaptive security on CPPS, proposed by Settanni et al

promptly reacting to cyber threats. The four phases of MAPE-K reference model in the CPPS are illustrated in Fig. 17, which consist of:

- Monitor: The result of this phase consists of the observed events by the sensors of the CPPS, which have to be inspected to identify security issues.
- Analyze: It analyses the acquired data through anomaly detection techniques, which gives information about the system's security status. Security metrics are examined, if they indicate a state of non-security, a change request is sent to Plan phase.
- Plan: It selects a strategy to mitigate the security issues from self-adaptation policies.
- Execute: It is responsible for executing the mitigation plan, using the available actuators on the CPPS.

4 Discussion

MAPE-K is a reference model that can be used to enable autonomic computing in different system types. This article centered on one of its applications, which is to achieve adaptive security. Ten approaches where surveyed, some of them focused on providing adaptiveness to a specific security objective or object, and others consisted of generic adaptive security systems. In Table 1 a comparison of the objectives which the models aim to make adaptive is presented. Four of the models centered on security objectives related to the management of resource access permissions, which are [9–11, 17]. Due to the complexity of predicting the user's behavior about resource access at design-time, adaptive computing can be used to dynamically adjust access permissions of a particular or group of users at runtime. Therefore, making the system more effective, efficient, and robust, because the system and environment current state, user needs, and system failures are taken into account when deciding which users are allowed to access resources, which of them, and in which manner. Also, the user's behavior can be used to reason if he/she is trustworthy; thus, access permissions can be granted or removed according to it, to prevent security attacks.

Table 1 Adaptation objective of the models

Model	Adaptation objective
[10]	Authorization
[11]	Access control
[9]	Access control
[12]	Security policies
[13]	Generic approach
[14]	Security policies
[15]	Security policies
[16]	Generic approach
[17]	Access control
[18]	Anomaly detection

Table 2 Target environment of the models

Model	Environment
[10]	Generic approach
[11]	Smart spaces
[9]	Cloud computing
[12]	WSN
[13]	Generic approach
[14]	Mobile devices
[15]	Smart camera networks
[16]	Intelligent vehicles
[17]	IoT
[18]	Cyber physical production systems

Considering that the environment where the model will be executing have an impact on the manner that systems interact, their security needs and vulnerabilities, and the system restrictions, some models were designed for the characteristics, limitations, requirements, and threats of a particular environment. In Table 2 the intended environment of the models is presented.

On MAPE-K model, the shared knowledge is an essential component because it allows the system to learn what is the best way to conduct itself in a particular situation. Due to this, each one of the models proposed an approach to knowledge creation. In Table 3 a comparison of the shared knowledge structure of the models is presented.

Table 3 Structure of the shared knowledge of the models

Model	Knowledge structure
[10]	Policy model and usage statistics
[11]	ISMO
[9]	Relationship degree between access control attributes
[12]	Generic approach
[13]	Service state, current state, adaptation and security goal
[14]	Service state, current state, adaptation and security goal
[15]	Service state, current state, adaptation and security goal
[16]	Predefined security rules, laws, and possible solutions on given circumstances
[17]	User profile, historical interaction data, and policy data
[18]	Generic approach

5 Conclusion

Information systems have evolved into interconnected heterogeneous components, which makes it difficult to anticipate their interactions at design-time. Also, as the security threats increase in number and complexity, static security measures are not always capable of confronting them in all the possible contexts. A solution could be autonomic computing, which consists of systems that can manage themselves at runtime following high-level objectives from the administrators. Thus, by using autonomic software, the security mechanisms can be adapted at runtime according to the environmental changes that impact on the required security level. A methodology for autonomic computing is MAPE-K reference model, which consists of a feedback loop that monitors environmental changes, analyses their impact on the high-level goals, and plans and executes actions to address them. Adaptive security models have used MAPE-K to dynamically adapt security measures on changing environments to met security objectives. In this manuscript, ten models were described and compared according to their adaptive security objective, the environment to which they were designed, and the structure of their knowledge base.

Funding This work was funded by CONACYT, under grant number 536467.

References

1. Kephart, J.O., Chess, D.M.: The vision of autonomic computing. Computer (2003). https://doi.org/10.1109/MC.2003.1160055
2. IBM: An architectural blueprint for autonomic computing. IBM White Paper. https://www-03.ibm.com/autonomic/pdfs/ACBlueprintWhitePaperV7.pdf (2005). Accessed 29 Dec 2018
3. Brun, Y., Serugendo, G.D.M., Gacek, C., Giese, H., Kienle, H., Litoiu, M., Müller, H., Pezzè, M., Shaw, M.: Engineering Self-Adaptive Systems Through Feedback Loops. Software Engineering for Self-Adaptive Systems, pp. 48–70 (2009). https://doi.org/10.1007/978-3-642-02161-9_3
4. D'Angelo, M., Caporuscio, M., Napolitano, A.: Model-driven engineering of decentralized control in cyber-physical systems. In: 2017 IEEE 2nd International Workshops on Foundations and Applications of Self* Systems (FAS*W) (2017). https://doi.org/10.1109/FAS-W.2017.113
5. D'Angelo, M., Napolitano, A., Caporuscio, M., Chess, D.M.: CyPhEF: a model-driven engineering framework for self-adaptive cyber-physical systems. In: 2018 IEEE/ACM 40th International Conference on Software Engineering: Companion (ICSE-Companion) (2018). ISSN 2574-1934
6. Eryilmaz, E., Trollmann, F., Albayrak, S.: Conceptual application of the MAPE-K feedback loop to opportunistic sensing. In: Sensor Data Fusion: Trends, Solutions, Applications (SDF) (2015). https://doi.org/10.1109/SDF.2015.7347697
7. Seo, Y., Kim, Y., Lee, E., Seol, K., Baik, D.: Design of a smart greenhouse system based on MAPE-K and ISO/IEC-11179. In: 2018 IEEE International Conference on Consumer Electronics (ICCE) (2018). https://doi.org/10.1109/ICCE.2018.8326276
8. Adler, R., Schneider, D., Trapp, M.: Development of safe and reliable embedded systems using dynamic adaptation. In: 2007 1st Workshop on Model-Driven Software Adaptation M-ADAPT'07 at ECOOP (2007). https://doi.org/10.1007/978-3-540-78195-0_13
9. Ma, S., Wang, Y.: Self-adaptive access control model based on feedback loop. In: 2013 International Conference on Cloud Computing and Big Data (2013). https://doi.org/10.1109/CLOUDCOM-ASIA.2013.94
10. Bailey, C., Chadwick, D.W., Lemos, R.D.: Self-adaptive authorization framework for policy based RBAC/ABAC models. In: 2011 IEEE Ninth International Conference on Dependable, Autonomic and Secure Computing (2011). https://doi.org/10.1109/DASC.2011.31
11. Evesti, A., Suomalainen, J., Ovaska, E.: Architecture and knowledge-driven self-adaptive security in smart space. Computer (2013). https://doi.org/10.3390/computers2010034
12. Pinto, M., Gámez, N., Fuentes, L., Amor, M., Horcas, J. M., Ayala, I.J.: Dynamic reconfiguration of security policies in wireless sensor networks. J. Sens. (2015). https://doi.org/10.3390/s150305251
13. Amoud, M., Roudies, O.: MaPE-K-based approach for security @ runtime. In: 2016 IEEE International Conference on Software Science, Technology and Engineering (SWSTE) (2016). https://doi.org/10.1109/SWSTE.2016.28
14. Amoud, M., Roudies, O.: Dynamic adaptation and reconfiguration of security in mobile devices. In: International Conference on Cyber Incident Response, Coordination, Containment Control (Cyber Incident) (2017). https://doi.org/10.1109/CYBERINCIDENT.2017.8054639
15. Amoud M., Roudies O.: Using combination of MAPE-K and DSPL to secure smart camera networks. In: Proceedings of the International Conference on Industrial Engineering and Operations Management (2017)
16. Singh, M., Kim, S.: Reconcile security requirements for intelligent vehicles. In: 2017 17th International Conference on Control, Automation and Systems (ICCAS) (2017). https://doi.org/10.23919/ICCAS.2017.8204251
17. Gwak, B., Cho, J., Lee, D., Son, H.: TARAS: trust-aware role-based access control system in public internet-of-things. In: 2018 17th IEEE International Conference on Trust, Security and Privacy in Computing and Communications. 12th IEEE International Conference on Big Data Science and Engineering (TrustCom/BigDataSE) (2018). https://doi.org/10.1109/TrustCom/BigDataSE.2018.00022

18. Settanni, G., Skopik, F., Karaj, A., Wurzenberger, M., Fiedler, R.: Protecting cyber physical production systems using anomaly detection to enable self-adaptation. In: 2018 IEEE Industrial Cyber-Physical Systems (ICPS) (2018). https://doi.org/10.1109/ICPHYS.2018.8387655
19. Muñndez, J.C., Mazo, R., Salinesi, C., Tamura, G.: 10 Challenges for the specification of self-adaptive software. In: 2018 12th International Conference on Research Challenges in Information Science (RCIS) (2018). https://doi.org/10.1109/RCIS.2018.8406640
20. Conti, M., Das, S.K., Bisdikian, C., Kumar, M., Ni, L.M., Passarella, A., Roussos, G., Tröster, G., Tsudik, G., Zambonelli, F.: Looking ahead in pervasive computing: challenges and opportunities in the era of cyber-physical convergence. In: Pervasive and Mobile Computing, Technology and Engineering (SWSTE) (2012). https://doi.org/10.1016/j.pmcj.2011.10.001
21. Elkhodary, A., Whittle, J.: A survey of approaches to adaptive application security. In: Proceedings of the 2007 International Workshop on Software Engineering for Adaptive and Self-Managing Systems (2007). https://doi.org/10.1109/SEAMS.2007.2
22. ANSI: American National Standard for Information Technology—Role Based Access Control (2012). INCITS 359-2012
23. ISO/IEC: Security Frameworks for Open Systems: Access Control Framework. ISO/IEC 10181-3 (1996)
24. Evesti, A., Savola, R., Ovaska, E., Kuusijärvi, J.: The design, instantiation, and usage of information security measuring ontology. In: Proceedings of the 2nd International Conference on Models and Ontology-Based Design of Protocols, Architectures and Services, IARIA (2011)
25. Hallsteinsen, S., Hinchey, M., Park, S., Schmid, K.: Dynamic software product lines. In: Systems and Software Variability Management: Concepts, Tools and Experiences (2013). https://doi.org/10.1007/978-3-642-36583-6_16
26. Hinchey, M., Park, S., Schmid, K.: Building dynamic software product lines. Computer (2012). https://doi.org/10.1109/MC.2012.332
27. INTER-TRUST European Project. https://github.com/Inter-Trust/. Accessed 8 Jan 2019
28. Horcas, J., Pinto, M., Fuentes, L.: Runtime enforcement of dynamic security policies. In: 2018 IEEE/ACM 40th International Software Architecture (2014). ISBN 978-3-319-09970-5
29. Gámez, N., Fuentes, L.: FamiWare: a family of event-based middleware for ambient intelligence. Pers. Ubiquitous Comput. (2011). https://doi.org/10.1007/s00779-010-0354-0
30. Haugen, Ø., Møller-Pedersen, B., Olsen, G.K., Svendsen, A.: Adding standardized variability to domain specific languages. In: 2008 12th International Software Product Line Conference (2008). https://doi.org/10.1109/SPLC.2008.25
31. SanMiguel, J.C., Micheloni, C., Shoop, K., Foresti, G.L., Cavallaro, A.: Self-reconfigurable smart camera networks. Computer (2014). https://doi.org/10.1109/MC.2014.133
32. Pinel, E.C., Long, A.E., Landau, M.J., Alexander, K., Pyszczynski, T.: Seeing I to I: a pathway to interpersonal connectedness. J. Pers. Soc. Psychol. (2006). https://doi.org/10.1037/0022-3514.90.2.243
33. Pinel, E.C., Long, A.E., Crimin, L.A.: I-sharing and a classic conformity paradigm. Soc. Cognit. (2010). https://doi.org/10.1521/soco.2010.28.3.277

Interpretable Machine Learning from Granular Computing Perspective

Raúl Navarro-Almanza, Juan R. Castro and Mauricio A. Sanchez

Abstract Machine Learning (ML) is a method that aims to learn from data to identify patterns and make predictions. Nowadays ML models have become ubiquitous, there are so many services that people use in their daily life, consequently, those systems affect in very ways to the final users. Recently, there is a special interest on the right of the final user to know why the system generates some output; this field is called Interpretable Machine Learning (IML). Granular Computing (GrC) paradigm is focused in knowledge modeling inspired by human thinking. In this work we conduct a survey of the state of the art in IML and GrC fields to settle the bases of the possible contribution of each other with aims to build more interpretable and accurately ML models.

Keywords Interpretable machine learning · Granular computing · Explainable artificial intelligence

1 Introduction

Nowadays the Artificial Intelligence is immerse in human daily life trough automated decision making systems. It is common to interact with those systems in such vary of services such as bank loans, recommender systems, medical context, and so on. The decision making by those systems impact directly to the final user due the possibility of service denegation.

R. Navarro-Almanza (✉) · J. R. Castro · M. A. Sanchez
Universidad Autónoma de Baja California, Calzada Universidad 14418, Tijuana, Baja California, Mexico
e-mail: rnavarro@uabc.edu.mx
URL: http://fcqi.tij.uabc.mx/

J. R. Castro
e-mail: jrcastror@uabc.edu.mx

M. A. Sanchez
e-mail: mauricio.sanchez@uabc.edu.mx

M. A. Sanchez et al. (eds.), *Applied Decision-Making*, Studies in Systems, Decision and Control 209, https://doi.org/10.1007/978-3-030-17985-4_8

Just as there are situations in which the impact of the decision is not critical, there may be situations where it affects the final user considerably. Because of this, it is necessary to know in depth how the models reach some conclusion and generate their outputs.

Interpretable Machine Learning area aims to create Machine Learning models easily understandable by humans. IML has been grow in popularity due the decision maked by those models can affect directly to customers, moreover, there are spreadly adopted by the enterprises to automatize certain process; there are so many services that use ML models involved in the daily human life.

Research questions:

1. What is interpretability in machine learning context?
2. Why is interpretability important in ML?
3. What is the state of the art in interpretability ML?
   (a) How the model interpretability is measured?
   (b) What are the perspectives to achieve the interpretability?
   (c) How can we classify the actual approaches in IML?

Learning models called black boxes have been widely adopted by both the scientific and business community to solve problems of very different natures, the common aspect between these applications is the objective of achieving the best possible performance (minimum error). However, it is vital to know the reasons why the learning model gives certain results, this in order to achieve confidence in the model. In black box learning models the data scientist does not know the behavior of the model when it is generated to solve a certain task, so this type of models are critical in problems such as medical diagnosis, detection of terrorism and other applications that suppose a risk for people [24].

There is a special attention in achieving that the models of automatic learning grant to the user explanations of the reasoning to arrive at a certain result, the name of the area that is in charge to solve this situation is commonly called Explainable Artificial Intelligence or Interpretable Machine Learning. In this work, the concepts will be mentioned indistinctly. Explanations are necessary for understanding the models and generating confidence in machine learning models. If there are explanations of the model, even if it presents an acceptable performance if the explanations do not correspond with the expert knowledge of the domain, the use of that particular model could easily be omitted; the most important thing is that it would offer the user the possibility of creating new models in order to avoid the errors detected through the explanations [26].

The machine learning models have been able to solve increasingly complex problems, however, these models have become more complex every time they are difficult to understand in the development of their conclusions. Due to this tendency an interest has arisen in the area of Explainable Artificial Intelligence, as indicated by the emergence of different government programs, such as DARPA's Explainable Artificial Intelligence program[1] and the General Regulation of Data Protection of the

[1]https://www.darpa.mil/program/explainable-artificial-intelligence.

European Union [9] where emphasis is placed on the user's right to an explanation, this made from the point of view of the end final user.

In the area of Explainable Artificial Intelligence there are several proposals to achieve the creation of explanations of machine learning models, some of them are focused on including the user in the learning process (human in the cycle) [9], unification of logic and probability to provide learning models with a formal representation [5], building knowledge bases from results generated by learning models supported by explicit knowledge of the domain expert [36].

One of the main problems in creating interpretable learning models lies in the form of representation of the data, for which there are several possibilities such as treating the data as fuzzy sets, probabilistic sets, rough sets, this in order to reduce the complexity of the data. The reduction of dimensionality of data, trough feature transformation or relevant feature detection, might benefits the tracking of the transformation of the data to later generate rules. This is one of the arguments by which it is feasible to use Granular Computation, whose paradigm is inspired by how the human mind process the information through multiple perspectives in hierarchy form. This offers the possibility of abstracting at different levels the data for the generation of different types of explanations, from the most specific to the most general.

The main process in Granular Computing (GrC) is to represent objects or patterns through units called granules, that process is called granulation. These granules can be composed of other granules. One of the main characteristics of the granules is that they can be represented by many kind of models such as those discussed above. The most important about this characteristic is that the granule can be modeled by a composition of many mathematical models, that is to say, can be modeled from different perspectives.

The main objective of this work is to analyze the contributions that could bring the granular computing paradigm to the interpretable machine learning area. The specific characteristics identified of GrC to potentially improve IML will be detailed in this work. The past problem description in GrC is linked to IML problem in terms of how to select the essential characteristics of the phenomena to build an explication easily understandable. The focus of this research is to detail the state of the art in IML and GrC as a survey to better understanding of the problem definition of interpretability from a GrC perspective. In a later sections it is describe in detailed form some of the open issues that, in some way, are shared in Interpretable Machine Learning and Granular Computing fields.

2 Interpretability in Machine Learning Context

Interpretability in machine learning context it is defined as the capability of some machine learning model be easily understandable by the data scientist and even the final user of the model, such as customers of some application. The are some definitions of interpretability from an non mathematical perspective identified by Molnar in [19], such as: Interpretability the degree to which human can understand the cause of a decision [18]. And, Interpretability is the degree to which a human can consistently predict the model's result [13].

The use of machine learning models, in daily human life, have becoming ubiquitous [29]. They are present in applications and services such as loans application, recommender systems for multimedia contents, on-line shopping and so many others. The decision making by those systems directly affect the final user, however, all of them have the right of know why some decision have been made; of course, in many cases machine learning model decisions does not impact negatively to the user in case of making some mistake, that means, those system are not critical.

The interpretability in machine learning is growing in importance because of critical systems whose decision could affect negatively to the user, such as medical applications [15], loan approvement systems, pedestrian detection in autonomous driving car, etc. The interpretability not only are focused in give to the final user an explication of why some decision have been made, but it might be give to the data scientist insights of how the ML model it is working to arise the conclusions to make predictions, and more importantly, fix undesired behavior of the system in the face of some rare input data (different distribution of data in training phase).

2.1 Machine Learning

Machine learning is and artificial intelligence sub-area that aims to automatically find the best parameters of some mathematical model to describe some phenomena. When there are many data measured that describe these phenomenon, it can be fin a function that approximate it behavior trough the samples, this specific case it is called supervised learning. Those models can be use to predict some quantity that is called regression task (which its output are real numbers), or to predict some class that is called classification task (which its output are natural numbers). The best parameters of some model are finding by optimization methods, for example, the well-known gradient descent that is very popular in Neural Networks models.

A machine leaning model can be represented as a function $f : X \times W \rightarrow Y$ where X and Y represent the inputs and outputs respectively, these data is contained in the training dataset; and W are the weights of that adjust the output of the system (in supervised training). The objective of learning methods is to find the best weights ($W^{\star}$) that satisfy the next relation $f(X, W^{\star}) \approx Y$. In that brief machine learning model description, it is noticeable that the main objective of these models is to approximate to the target value, diminishing the importance of how and why of the decision is generated.

2.2 Interpretable Machine Learning Models

In machine learning domain there are a huge amount of different mathematical models to represent some phenomena. In terms of interpretability exists a consensus of which models are easily interpretable (or simply interpretable). There is a common idea that the simpler the more interpretable model is, that sounds logic if we consider the fewer parameters and generally the lack of non-lineal transformations.

The are some fields that already have approaches to get better understanding of the problem and solution through IML, such as e-learning to help students to better understanding of subjects [32]; in biology to understanding high-order interactions drive gene expression [3]; in medicine to build a system to predict patients mortality with ST-elevation myocardial infarction [15], building interpretable decision tree for diagnosis and other tasks [27]; in neuroscience to studied a family of N-back working memory [7]; in vision-based systems to achieve better interpretability by dictionary of atomic shapes in decision boundaries [28]. In text analytics to document categorization as a classification task with individual word relevance [1]; in robotics to generate explanations in humanoid robots [4, 12]. In Fig. 1 is shown the tendency in the number of publications on Interpretable Machine Learning from 1989 to 2018, these information was acquired from Scopus database.

In machine learning problems there is a common threshold between the quality of *what* is the predicted result of the model and *why* the model gives an specific output. The main problem is that apparently to improve one of them affects negatively the another one.

The machine learning models can be classified by their transparency, in this work, this concept refers to the ease with which the user can understand why the model generated a certain result. There are two where know classes of models: gray box model and black box models, which will be detailed below.

The machine learning models that are considered has gray box models, they stand out for their lack of non-linear computations, such as decision tree and linear models.

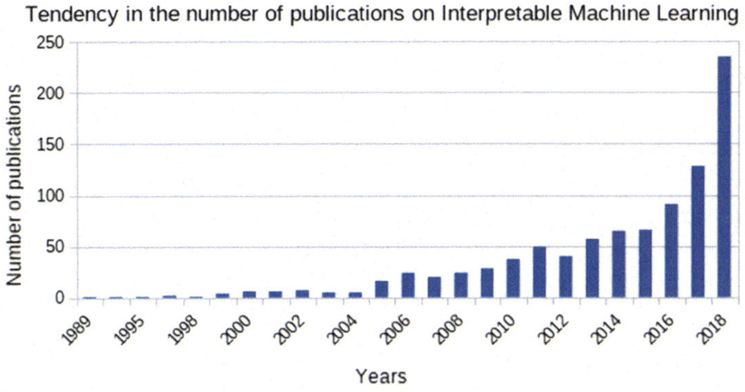

Fig. 1 Tendency in the number of publications on interpretable machine learning

In those models can be performed some operations to achieve some understanding of how the system make the predictions. Also can be explained by contrafactual cases, that means to know which features and how can be modified in order to change its output to some desired result of the model.

The black box models generally has a huge amount of hyper-parameters and weights, same for the non-linear computations that perform. The most popular black box models are the Neural Networks, due of the large amount of successive non-linear transformations. It is well know the lack of interpretability of these models, however, they are capable to achieve better behavior in so many complex task.

2.3 Scope of Interpretability

There are different perspectives in terms of model interpretability, one of them is focused on generate an explanation for each instances that is evaluated by the model. On the other hand, there is an perspective that aims to explain the entire behavior of the model. Both trends have their own advantages and drawbacks, the selection of one of them depends on the application domain and goals.

Global interpretability aims to generate an explanation of the entire solution surface of the model. The most common artifact to create such explanation relies in rule-based models. These approaches consist of creating a knowledge base of the entire domain of the problem from the solution surface, within the most outstanding works it is found that of Lakkaraju et al. [14], where the knowledge base is composed of decision sets and in the learning process various objectives are taken into account such as fidelity, disambiguation, interpretability.

Local interpretability is focused on generate an explanation for each prediction (or evaluation). Usually for this kind of explanation relies on identify the most important attributes for some model output. There some approaches where its interpretability relies in the measure of certainty to give to the user of machine learning model [30].

2.4 Interpretability Models

Another perspective of model explanations depend by the model itself, there are techniques of interpretability that only can be used in certain machine learning models, such as Learning Vector Quantization model [6, 11], Multi-operator Temporal Decision Trees [25], Bayesian-based model for learning rule sets [31]. In the other hand, there are techniques that does not require some specific model to work. Generally, the explanation techniques that require a specific model are more interpretable, because of the understanding of model behavior, however, the model-agnostic techniques can be used in wider application domain.

Model-dependent interpretability techniques depends of the machine learning model that is used, the insights in those techniques are more easily understandable by the data scientist. Conversely, In model-agnostic interpretable techniques, the selected model for pattern discovery are irrelevant, a common way to understand them is to visualize the model as a black box in where only their inputs and outputs are analyzed.

3 Granular Computing

The Granular Computing paradigm is inspired by how the human mind process the information through multiple perspectives in hierarchy form. This hierarchy thinking form allows to focus in different essential attributes inherent of some object, depending of the requirements in reasoning processes.

Granular computing is a multidisciplinary area that encompasses theories, techniques, models and methodologies to solve complex problems. It is a paradigm of solving problems with a single basic element called a granule, which are formulated through the abstraction of properties in common between objects or patterns in the data. The properties that can exist between these objects or patterns are similarity, equality, proximity [21]. Granular Computing is growing in popularity as can be show in Fig. 2, where the number of publications about Granular Computing is plotted trough the years.

In granular computation, the process of creation of the granules is crucial and can be represented by various figures and sizes, as well as the interactions between them. Granular representation is especially useful in situations where the problem presents uncertainty and incomplete data. A granule can be modeled as $\mathbf{A} : X \rightarrow \mathcal{G}(X)$ where X is the universe of discourse and $\mathcal{G}$ is the framework of information granules. The process of granulation it can be represented as a mapping from one domain to another.

There is an important concept in Granular Computing that is called Granular World (**G**) that refers to the environment of information granules that supports all process

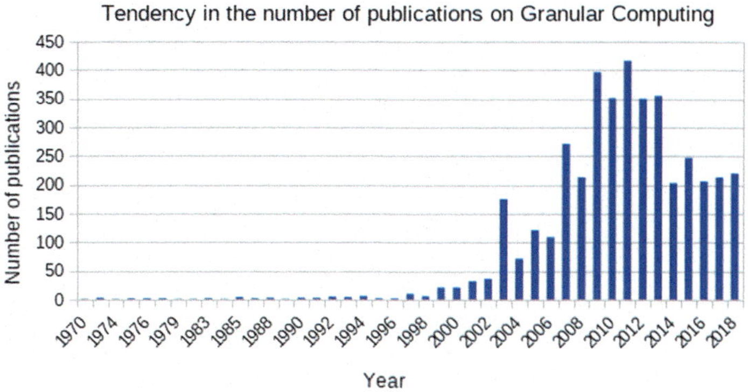

Fig. 2 Tendency in the number of publications on granular computing

in information granulation, information processing and information exchange [2]. In this environment can be composed by different granules order but same domain, each order has a different perspective of data, in other words, different hierarchies of granules; where the formal representation of Granular World is $\mathbf{G} = <X, \mathcal{G}, \mathbf{A}, \mathbf{C}>$, where

- $\mathbf{A}$: is a granule of information.
- $\mathbf{G}^n$: is the granular world.
  - X: universe of data discourse.
  - $\mathcal{G}(X)$: denotes the *framework* of granules of information.
  - $\mathbf{A}^n$: is a information granule (of type n).
  - $\mathbf{C}$: is a means of communication

Some interesting concepts in Granular Computing paradigm, such as (i) principle of justifiable granularity, (ii) coverage, and (iii) specificity, the first one is referred to a fundamental concept in the paradigm and is about forming information granule from data by available evidence. The concept of specificity refers to the level of characterization of a granule to detail some abstraction of the information granule, finally, coverage is referred to the amount of elements of the domain that correspond to some description of a information granule. There is a relationship between specificity and coverage, while the specificity increases the coverage decreases, and vice versa.

In Fig. 3 is shown the behavior of the coverage and specificity, if the specificity is very high then the coverage, that is to say, the number of elements in the universe of discourse that belong to the given specification. Otherwise, to achieve a high

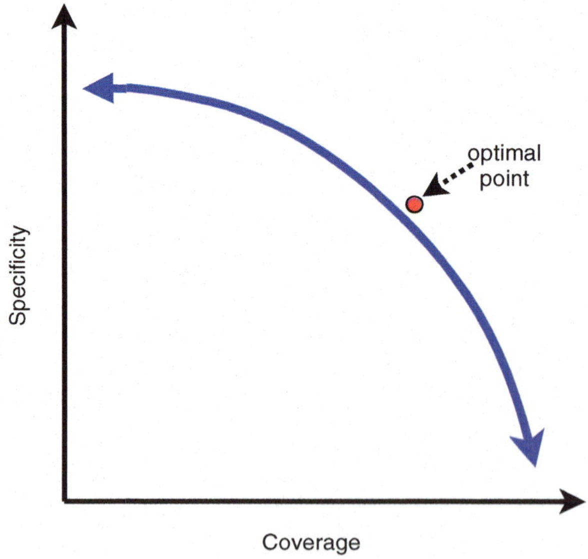

Fig. 3 Relationship between coverage and specificity in granular models

coverage it is necessary to reduce the specificity, that is, describe in very general form the information granules.

3.1 Granular Computing Models

A granule of information can be a class of numbers, group of images, a class of regions, a set of concepts, a set of objects and a category of data. The granules of information can be represented in the form of fuzzy sets, rough sets, fuzzy rough sets, neural networks, probabilistic sets, etc. [2]. Each representation model represents a different perspective of data, in addition, the composition in the granules is possible, therefore a granule of a certain order can be used to grant an explanation considering different points of view or several explanations according to the perspective.

There are some approaches that combine both worlds, machine learning and granular computing to overcome some issues related to the very large amount of data [10, 35], related to interpretability issues [20, 33], for knowledge representation as in Pedrycz [23] proposes the use of granules of information to represent a knowledge base using fuzzy sets, this knowledge base is created from different learning models.

In the following it is described some of the most used models in the area of Granular Computing.

Rough Sets

In the rough sets, situations are modeled where one can not distinguish the belonging of an object in a given set, but an element can be related to other elements, so upper and lower limits are defined for the description of the element in a given set, as shown below:

$$\underline{R}X = \bigcup\{Y \in U/R|Y \subseteq X\}$$

$$\overline{R}X = \bigcup\{Y \in U/R|Y \cap X \neq\}$$

where X is a classic set of objects, in a universe of speech U, R is the relation between existing elements in U. This description of elements is usually used to represent relationships between them.

Fuzzy Sets

Theory introduced by Zadeh [34] to model the uncertainty in natural language. It is a model to describe the degree of belonging in continuous space $A = \{(x, \mu_A(X))|x \in X\}$, where $\mu_A : X \rightarrow M$ is the membership function of x in A that maps X to the M membership space. The membership $\mu_A(X)$ indicates the degree of similarity (compatibility) of a x object to a concept characterized by the fuzzy set A. The domain of M corresponds to the range [0, 1].

One of the main characteristics of the modeling of granules by fuzzy sets is the one that can relate a granule to a linguistic variable and thus provide a higher degree of interpretability for the data scientist.

Fuzzy Rough Sets

In fuzzy rough sets, the aim is not only to model uncertainty, as in the case of fuzzy sets, but also to add concepts of incompleteness and imprecision. Here's how to model this type of sets:

$$(R_B \downarrow R_A)(x) = \inf_{y \in U} \max\{1 - R_B(x, y), R_A(x)\}$$

$$(R_B \uparrow R_A)(x) = \sup_{y \in U} \min\{1 - R_B(x, y), R_A(x)\}$$

where the upper and lower approximations involve fuzzy equivalence and decision classes.

Neural Networks

Granular neural networks are models capable of processing granular data, which may well come from both numeric and linguistic data [8]. This processing may consist of prediction of new information, extraction of information in the form of granules, fusion of sets of information granules or compression thereof.

The use of neural networks for processing or identification of granules is widely used, the most prominent works for generation of rules using granular neurons [8], neural networks using a scheme of functional neurons that allow various representations of granules [16], neuro-fuzzy networks for generating rules by granular computation [22].

4 Conclusions and Discussion

As is mentioned before in machine learning problems there is a common threshold between the quality of *what* is the predicted result of the model and *why* the model gives an specific output. The main problem is that apparently to improve one of them affects negatively the another one. This phenomena can be analogous to the behavior of information granules in terms of coverage and specificity, there is a relationship between specificity and coverage, while the specificity increases the coverage decreases, and vice versa.

The analogy raised above is the main justification for the possible contribution of GrC to the IML field. Because the explanations of the models can vary greatly with respect to how many attributes, decision sets, types of operators to consider. A very precise explanation, which contemplates several or all of the relevant factors discovered by the model, may not be of much help to a person not expert in the domain. Is quite possibly for a person not expert in the domain, an explanation that involves general terms, without so much detail about the phenomenon, could be more useful. Due to the ability to change the level of hierarchy of granules of information, the possible contribution of these is raised to achieve an ad hoc explanation to the user of the model, based on various factors such as user expertise, if the application is critical, if an object (instance) is anomalous or belongs to another data distribution and even to understand the model itself.

The GrC framework brings the possibility of create hybrid knowledge representation model composed by another mathematical models, that capture different perspectives of the data. Moreover, this framework has the characteristic of build nested models that can be use to represent different levels of abstraction, that might derive in multiple kind of explanations according to the needs of the user, either data scientist and final user (such as a customer).

Due the growing popularity of IML it is important to define how the approaches of building interpretable models (either model dependent or model-agnostic) should be evaluated by their performance, as is used in nowadays, furthermore, it is equivalently important to evaluate their interpretability in various ways according to exposed in these works [17, 26]:

- Measure how well the final user comprehend the decision of the model.
- Measure how easily the data scientist can understand why the model raise some conclusion.
- Measure the level of abstraction in which the explanation is made.
- Measure the knowledge gain of the data scientist while the using of interpretable machine learning model.

Acknowledgements This research was partially supported by MyDCI (Maestría y Doctorado en Ciencias e Ingeniería).

References

1. Arras, L., et al.: "What is relevant in a text document?": an interpretable machine learning approach. PLoS ONE **12**(8) (2017). ISSN 19326203. https://doi.org/10.1371/journal.pone.0181142
2. Bargiela, A., Pedrycz, W.: Granular Computing (2003). ISBN 978-1-4613-5361-4. https://doi.org/10.1007/978-1-4615-1033-8
3. Basu, S., et al.: Iterative random forests to discover predictive and stable high-order interactions. Proc. Natl. Acad. Sci. U.S.A. **115**(8), 1943–1948 (2018). ISSN 00278424. https://doi.org/10.1073/pnas.1711236115
4. Beaton, B.: Crucial answers about humanoid capital. In: ACM/IEEE International Conference on Human-Robot Interaction, pp. 5–12. IEEE Computer Society (2018). ISBN 9781450356152. https://doi.org/10.1145/3173386.3173391
5. Belle, V.: Logic meets probability: towards explainable AI systems for uncertain worlds. In: IJCAI International Joint Conference on Artificial Intelligence, pp. 5116–5120 (2017). ISSN 10450823. https://doi.org/10.24963/ijcai.2017/733
6. Brinkrolf, J., Hammer, B.: Interpretable machine learning with reject option. At-Automatisierungstechnik **66**(4), 283–290 (2018). ISSN 01782312. https://doi.org/10.1515/auto-2017-0123
7. Caywood, M.S., et al.: Gaussian process regression for predictive but interpretable machine learning models: an example of predicting mental workload across tasks. Front. Hum. Neurosci. **10** (2017). ISSN 16625161. https://doi.org/10.3389/fnhum.2016.00647

8. Ding, S., et al.: Granular neural networks. Artif. Intell. Rev. **1**(3), 373–384 (2014). ISSN 02692821. https://doi.org/10.1007/s10462-012-9313-7
9. Goodman, B., Flaxman, S.: European Union regulations on algorithmic decision-making and a "right to explanation", pp. 1–9 (2016). ISSN 0738-4602. https://doi.org/10.1609/aimag.v38i3.2741. arXiv: 1606.08813
10. Guo, H., Wang, W.: Granular support vector machine: a review. Artif. Intell. Rev. **51**(1), 19–32 (2019). ISSN 15737462. https://doi.org/10.1007/s10462-017-9555-5
11. Hofmann, D., et al.: Learning interpretable kernelized prototype-based models. Neurocomputing **141**, 84–96 (2014). ISSN 09252312. https://doi.org/10.1016/j.neucom.2014.03.003
12. Huang, S.H., et al.: Enabling robots to communicate their objectives. Auton. Robots 1–18 (2018). ISSN 09295593. https://doi.org/10.1007/s10514-018-9771-0
13. Kim, B., Khanna, R., Koyejo, O.O.: Examples are not enough, learn to criticize! Criticism for interpretability. In: Lee, D.D., et al. (eds.) Advances in Neural Information Processing Systems, vol. 29, pp. 2280–2288. Curran Associates, Inc. (2016)
14. Lakkaraju, H., Bach, S.H., Leskovec, J.: Interpretable decision sets: a joint framework for description and prediction. In: Proceedings of the ACM SIGKDD International Conference on Knowledge Discovery and Data Mining, 13–17 August, pp. 1675–1684. Association for Computing Machinery (2016). ISBN 9781450342322. https://doi.org/10.1145/2939672.2939874
15. Li, X., et al.: Using machine learning models to predict in-hospital mortality for ST-elevation myocardial infarction patients. In: Marie-Christine, J., Dong-sheng, Z., Gundlapalli, A.V. (eds.) Studies in Health Technology and Informatics, vol. 245, pp. 476–480 (2017). ISSN 18798365. https://doi.org/10.3233/978-1-61499-830-3-476
16. Loia, V., Tomasiello, S.: Granularity into functional networks. In: 2017 3rd IEEE International Conference on Cybernetics, CYB-CONF 2017—Proceedings (2017). https://doi.org/10.1109/CYBConf.2017.7985781
17. Mencar, C., Fanelli, A.M.: Interpretability constraints for fuzzy information granulation. Inf. Sci. **178**(24), 4585–4618 (2008). ISSN 00200255. https://doi.org/10.1016/j.ins.2008.08.015
18. Miller, T.: Explanation in artificial intelligence: insights from the social sciences (2017). arXiv: 1706.07269
19. Molnar, C.: Interpretable Machine Learning. https://christophm.github.io/interpretable-ml-book/ (2019)
20. Nápoles, G., et al.: Fuzzy-rough cognitive networks. Neural Netw. **97**, 19–27 (2018). ISSN 18792782. https://doi.org/10.1016/j.neunet.2017.08.007
21. Pal, S.K., Ray, S.S., Ganivada, A.: Granular Neural Networks, Pattern Recognition and Bioinformatics (2010). ISBN: 9783319571133
22. Panoutsos, G., Mahfouf, M.: A neural-fuzzy modelling frame-work based on granular computing: concepts and applications. Fuzzy Sets Syst. **161**(21), 2808–2830 (2010). ISSN 01650114. https://doi.org/10.1016/j.fss.2010.06.004
23. Pedrycz, W., Chen, S.-M.: Granular Computing and Intelligent Systems, p. 305 (2011). ISBN 9783642017988. https://doi.org/10.1007/978-3-642-19820-5
24. Ribeiro, M.T., Singh, S., Guestrin, C.: "Why should I trust you?": explaining the predictions of any classifier (2016). ISSN 9781450321389. https://doi.org/10.1145/1235. arXiv: 1602.04938
25. Shalaeva, V., et al.: Multi-operator decision trees for explainable time-series classification. In: Verdegay, J.L., Pelta, D.A., Yager, R.R., Bouchon-Meunier, B., Medina, J., Ojeda-Aciego, M., Cabrera, I.P. (eds.) Communications in Computer and Information Science, vol. 853, pp. 86–99 (2018). ISSN 18650929. https://doi.org/10.1007/978-3-319-91473-2_8
26. Smith, A., Nolan, J.J.: The problem of explanations with-out user feedback. In: CEUR Workshop Proceedings, vol. 2068 (2018). ISSN 16130073
27. Valdes, G., et al.: MediBoost: a patient stratification tool for interpretable decision making in the era of precision medicine. Sci. Rep. **6** (2016). ISSN 20452322. https://doi.org/10.1038/srep37854
28. Varshney, K.R.: Interpretable machine learning via convex cardinal shape composition. In: 54th Annual Allerton Conference on Communication, Control, and Computing, Allerton 2016, pp. 327–330. Institute of Electrical and Electronics Engineers Inc. (2017). ISBN 9781509045495. https://doi.org/10.1109/ALLERTON.2016.7852248

29. Varshney, K.R.: Engineering safety in machine learning (2016). arXiv:1601.04126 [stat.ML]
30. van der Waa, J., et al.: ICM: an intuitive model independent and accurate certainty measure for machine learning. In: Rocha, A.P., van den Herik, J. (eds.) ICAART 2018—Proceedings of the 10th International Conference on Agents and Artificial Intelligence, vol. 2, pp. 314–321. SciTePress (2018). ISBN 9789897582752
31. Wang, T., et al.: Bayesian rule sets for interpretable classification. In: Baeza-Yates, R., Domingo-Ferrer, J., Zhou, Z.-H., Bonchi, F., Wu, X. (eds.) Proceedings—IEEE International Conference on Data Mining, ICDM, pp. 1269–1274. Institute of Electrical and Electronics Engineers Inc. (2017). ISBN 9781509054725. https://doi.org/10.1109/ICDM.2016.130
32. Williams, J.J., et al.: Enhancing online problems through instructor-centered tools for randomized experiments. In: Conference on Human Factors in Computing Systems—Proceedings, Apr 2018. Association for Computing Machinery (2018). ISBN 9781450356206; 9781450356213. https://doi.org/10.1145/3173574.3173781
33. Xu, X., et al.: A new method for constructing granular neural networks based on rule extraction and extreme learning machine. Pattern Recognit. Lett. **67**, 138–144 (2015). ISSN 01678655. https://doi.org/10.1016/j.patrec.2015.05.006
34. Zadeh, L.A.: Fuzzy sets. Inf. Control **8**(3), 338–353 (1965). ISSN 0019-9958. https://doi.org/10.1016/S0019-9958(65)90241-X
35. Zhu, X., Pedrycz, W., Li, Z.: Granular representation of data: a design of families of ϵ-information granules. IEEE Trans. Fuzzy Syst. **26**(4), 2107–2119 (2018). ISSN 10636706. https://doi.org/10.1109/TFUZZ.2017.2763122
36. Zhuang, Y.-t., et al.: Challenges and opportunities: from big data to knowledge in AI 2.0. Front. Inf. Technol. Electron. Eng. **18**(1), 3–14 (2017). ISSN 2095-9184. https://doi.org/10.1631/FITEE.1601883

Nonlinear Modeling: Lessons Learned and Room for Improvement in the M&S Literature

E. Dante Suarez

Abstract The field of Modeling and Simulation (M&S) has been growing at a rapid pace in the last few decades, particularly since the integration of growing computational capabilities, and the incorporation of more and more researchers coming from both computational as well as more traditional social, physical and biological sciences. At the same time, the ideas encapsulated in the growing paradigm of nonlinearity and Complexity have been gaining traction and have become increasingly accepted in the Literatures of all fields of study. In this sense, the rise of complexity science and of computer simulations have happened simultaneously because they have co-evolved and found inspiration from each other. This essay attempts to serve as a type of mirror in which both Complexity and Simulation can look at each other, understand each other, and compare notes. Nonetheless, the point of the article is that this is not yet a perfect mirror, and that the state-of-the-art theoretical Literature on complexity and computer simulations still do not yet see eye to eye. There are certainly lessons to be learned from the simulation side of the equation that have not been fully grasped and generalized. This work, however, focuses on the other side of the coin: On the theoretical insights that have not been appropriately modeled as of now, and about ways in which we can move the M&S field in that direction, with current and potential real-world applications of this increased rapport between theory and application.

1 Introduction

The vast and growing amount of computing power has opened research avenues that were impossible a few years ago, as well as forced researchers to reconsider the basic premises of the reductionist paradigm on which traditional linear sciences—such as neoclassical economics and individual selection theory in evolutionary biology—are built. Traditional linear science had been forced historically to assume away all

E. Dante Suarez (✉)
Department of Finance and Decision Sciences, School of Business, Trinity University, One Trinity Place, San Antonio, TX 78212, USA
e-mail: esuarez@trinity.edu

M. A. Sanchez et al. (eds.), *Applied Decision-Making*, Studies in Systems, Decision and Control 209, https://doi.org/10.1007/978-3-030-17985-4_9

interactions between the components of a system, such as the buyers and sellers in the stock market or the ants in an anthill; not because they were considered irrelevant, but because studying them was impossible. This is because the mathematics necessary to explain the whole system and its interactions quickly become intractable [1]. The answer to the conundrum was to find systems that could be described without considering the interaction between the component agents, and this gave rise to disciplines such as the highly mathematized neoclassical economics of the 20th century, in which culture and society do not exist. This essay uses the way that the field of economics has been developed in the last century, and how it is now usually taught (particularly at an undergraduate level), as a representative case study. The essential insight of these ideas can be applied to most other fields of scientific inquiry.

There is a joke about a drunken man looking for his keys under a streetlight one night, when a neighbor passes by and asks if he can help. "Where exactly did you lose your keys," the helpful neighbor asks. "Over there," answers the drunken man, pointing away into the darkness. "Then why are you looking for your keys here?" asks the neighbor, bewildered. "Because there is light here," the drunk replies matter-of-factly. Similarly, a linear science such as neoclassical economics had to assume away all agent interaction to be able to build a paradigm where the aggregate, macroeconomic behavior grew directly from mathematical models that understood the microeconomic behavior of such atomic agents as consumers and firms. In doing so, one of the most important aspects of the system, the structure in which the agents do interact, is completely erased. A linear model of reality that does not consider interaction between the forming parts will fail to adequately describe it, for it will lack an understanding of the structure of the system.

The intrinsic interaction and resulting structure of the system are often the actual phenomena that needs studying, or the components that create the observed behavior of interest, such as when we analyze the cohesion of an army unit, the coordination of the best football team, the culture of an enterprise, or the checks and balances of a government. Naturally, responses to this approach do exist in the Literature, and we can find a growing number of bustling areas and interdisciplinary fields such as, for example, econophysics, behavioral finance, and institutional economics. These fields, however, can represent smaller areas not yet accepted in the main paradigm of the broader field, may not often be currently taught at the undergraduate level, or may be upon closer inspection still grounded in a linear foundation.

Repeated interaction gives rise to structures that deserve attention, and can become much more important than the atoms, parts, components or agents that the system is built out of. Aggregates are therefore more than the sum of their parts, and we refer to such aggregates as emergent. To understand them, we cannot simply reduce the system's aggregate behavior to its minimal components, study them in isolation, and then rebuilt the system by multiplying the behavior of the average piece by the number of components in the system. In the emergent, nonlinear, non-reducible world of complexity, nature exists in many non-orthogonal levels, with each level potentially being governed by different laws, granularities and structure [2]. If we believe in that proposition, then we should expect to find a world full of emergent phenomena, with distinctive levels of interaction that have agency of their own [1].

In John Conway's Game of Life, for example, three simple rules that describe the behavior of simplified cells on a grid give rise to unexpected aggregate behavior, such as that of a so-called glider, which is a continuous pattern that moves across the screen, potentially carrying information to other areas of the grid [3]. If the reader has never seen this game, it may be a good idea to take a quick look at it, as the level of complexity that it creates is truly amazing. Several free programs are available on the internet, and one can quickly become familiarized with it. In this so-called game of life, the appearance of the glider is in no way specified in the original three rules of the system and can thus be considered to be an emergent property. But can we truly say that the glider was not implicitly coded into the three original rules? Is emergence simply another way of referring to our subjective lack of complete understanding of the ultimate product of the interaction of three simple rules? Could a more intelligent being see these rules and immediately recognize that they spell out 'glider'? The truth is that the concept of emergence, just as the general concept of complexity, has not fully been defined in a way that can be universally accepted, or that encapsulates all of the aspects and uses of the concepts.

Pessa [4] and more recently Tolk [1] detangle the concept of emergence and separate it into two categories: epistemological and ontological. Epistemological emergence refers to situations where a system has rules that can be understood, but it is irreducible to one single level of description for conceptual reasons. Think of the impossibility trying to type an essay in your computer all in zeroes and ones or describing an economy as a collection of protons and electrons. A system may therefore be epistemologically emergent if it follows rules that in principle exist and are generalizable, but that is bound by unpredictability because humans are not able to fully grasp these laws in their entirety in one single level of description. Pessa [4] points out that emergence can only be defined subjectively and in relation to a given observer. This means that what may be emergent to someone may be an obvious and direct result of the interacting parts of the system. Tolk [1] stresses the fact that computer simulations of these systems can only produce this type of epistemological emergence, as anything that is coded in the computer can be translated into knowable and reproducible mathematical functions.

Crutchfield [5], on the other hand, defines intrinsic emergence, which is objective and thus independent of the external observer. In this case, the emergent properties are truly novel and cannot be explained solely by the components and their interactions. Take biological life, for example: even if we understand all the physical and chemical laws that underpin it, we cannot say that we truly understand it, much less that we can recreate it. Following Tolk [1], we refer to this type of true emergence as ontological.

In summary, the linear paradigm prevalent across many scientific disciplines until the later part of the 20th century was chosen by necessity. In it, complex phenomena can be reduced, that is, analyzed by looking at the behavior of the average agent and then building the aggregate environment back up, disregarding in the process any structure or topology potentially created by the interaction between the (potentially adaptable) parts in the system. In a linear system, any variable can be parsed out and can be studied independently, the way a multivariate linear regression can dissect a dataset and express it as an addition of variables, linear coefficients, and random errors. The nonlinear view of reality, on the other hand, says that such reductionism, simplified analysis, and partitioning are not possible. Summarizing this dichotomy in simple words, imagine that when our shoulder hurts, a western, linear physician will look at our shoulder to cure it, and nothing more. Representing nonlinearity, a doctor steeped in the Eastern traditional medicine way of curing may need to look at the whole body to see what is wrong with its "chi," or life force.

This essay is centered on presenting how the field of Modeling and Simulation (M&S) has captured this increased focus on the interaction and structure. In other words, this work attempts to identify social and environmental phenomena, as well as other aspects of our modern-day world that the nonlinear paradigm has identified, to see if the M&S Literature has developed the right models, tools and methodologies to represent it. The main intention of this task is to identify those aspects of reality that remain fairly un-modeled.

For the most part, to tackle the novel challenges and opportunities that a nonlinear view of the world provides, the M&S field has relied heavily on Multi-Agent Systems (MAS)—a growing computational field that is able to capture many of the emergent social phenomena that researchers are interested in exploring. However, the problem with the MAS approach is that it is ultimately also plagued by some of the constraining assumptions of science's linear past. In particular, this essay is built on the idea that the definition of agents and agency is outdated, for it does not allow for the incorporation of a fractal, hierarchical and multidimensional view of the world that the complexity paradigm has revealed.

In contrast, we argue that such a view incorrectly de-contextualizes organisms that are partly independent agents, composed of relatively independent or modular organs, and partly belonging to groups and co-evolutionary processes that influence their decisions. Consider, for example, an idea that is prevalent in evolutionary biology: The Evolutionary Stable Strategy (ESS). This important concept is an expansion of the idea of a Nash equilibrium, and it refers to a strategy that if adopted by a population in a given environment, is impenetrable, meaning that it cannot be invaded by any alternative strategy that is initially rare. In relation to the original concept of a Nash equilibrium, ESS refers to a moment when all the individuals of a population are doing what is in their best interest, given what others are doing. The field of biology has many applications of this idea, including, for example, an original and seminal article by Taylor and Jonker [6], describing the concept of an ESS; a more recent application by Goldstein and Young [7], to the game of baseball; or a most recent and very interesting application to economics by Lindgren [8]. In a linear world where only individual organisms exist and evolution only acts at this level, we would expect ESS to be the prevalent norm across the animal kingdom. If, on the other hand, many more such levels other than that of the individual exist, then we may expect a lot of animals to display non-ESS, as they would be acting not as individuals, but rather as parts of larger systems that cannot be decontextualized.

We expect the world to be full of organisms following evolutionary unstable strategies, precisely because their agency is not absolute and reducible, but rather contextualized. In simple terms, the world is full of 'cooperating prisoners' such as those who cooperate in a prisoner's dilemma setting [9]. It is important to stress that, for the purpose of understanding behavior, what is considered the self—or the force of individuality—is not necessarily what the physical self is, but only that which is dearest. Parents may willingly give their lives for their children, just as a presidential guard is trained to take a bullet. What one considers dearest is the result of a complex process, making the procedure for defining what is the ultimate self potentially unfeasible. In the case of the guard, for example, it seems that the abstract concept of honor takes precedent.

Moreover, if we redefine collections of individuals as operational units—such as ants forming an indivisible anthill—then we must redefine what we mean by ubiquitous concepts such as altruism and selfishness [10]. We propose in this chapter to follow a benchmark position in which all actions are revealed as selfish once one understands what the benefited acting agent is. Traditionally, we began with a clearly defined agent and tried to understand its actions as a maximization of objectives given constraints. In the proposed paradigm, we assume maximization occurs, and then work towards the delineation of the benefited entity involved. Such an understanding of an individual as an agglomeration of relatively independent aims can serve as a starting point for a broader theory that describes the formation of complex hierarchical objects [11], stressing the in-group vs. out-group dichotomy as it applies to evolutionary biology, but also to economics, sociology, social psychology and political science.

2 What Complexity Teaches Us

In this section we lay out some of the most important concepts that complexity science has brought to the forefront of scientific discussion, providing a simple and condensed resource, especially for those new to these theoretical ideas. To provide an intuitive definition, we originally rely on Wikipedia. The purpose of this is to provide what is arguably the currently most common definition available, to then explain it in simple terms, expanded and nuanced, or adapted to our current discussion. The original definition from Wikipedia, as well as auxiliary concepts used, appear in italics.

2.1 Nonlinear System/Nonlinear Dynamics/Nonlinearity

In mathematics and science, a nonlinear system is a system in which the change of the output is not proportional to the change of the input. Nonlinear problems are of interest to engineers, biologists, physicists, mathematicians, and many other scientists because most systems are inherently nonlinear in nature. Nonlinear dynamical systems, describing changes in variables over time, may appear chaotic, unpredictable, or counterintuitive, contrasting with much simpler linear systems.

This, we would argue, is an older and perhaps misleading definition of nonlinearity. As we use it in complexity, the concept of emergence refers to the idea that the aggregate is completely different than the sum of its parts, such as in the way that a mind is much more than a collection of neurons. The mind exists in a completely different dimension than the physical, chemical, and biological world of the neurons and the brain. In this sense, when we speak of nonlinearity in our field, we are often not speaking about non-proportionality, in the way that two plus two equals five, but rather we are referencing emergence and the different dimensions or levels of reality that this creates, and as such, two plus two might equal blue.

2.2 Complexity

Complexity characterizes the behavior of a system or model whose components interact in multiple ways and follow local rules, meaning there is no reasonable higher instruction to define the various possible interactions. The term is generally used to characterize something with many parts where those parts interact with each other in multiple ways, culminating in a higher order of emergence greater than the sum of its parts. The study of these complex linkages at various scales is the main goal of complex systems theory.

Complexity is a difficult beast to tame. To this day, there is no definition commonly accepted by all scientists. This is at least in part, we would argue, because those of us working in the field recognize that it is a nascent paradigm with room to grow, so that we must be careful to be inclusive, and not close the door prematurely on concepts, methodologies or phenomena that deserve study. The definition provided captures some of the most important features that are often expressed, such as many interacting and potentially adaptive parts, no centralized control (which makes many of the systems of complexity bottom-up, that is, with aggregates that are created from the micro to the macro, and not the other way around), and emergent aggregate behavior.

2.3 Chaos Theory

Chaos theory is a branch of mathematics focusing on the behavior of dynamical systems that are highly sensitive to initial conditions.

As summarized by Edward Lorenz, who famously discovered these types of nonlinear dynamics while modeling the weather: "*Chaos happens when the present determines the future, but the approximate present does not approximately determine the future.*"

Chaos and Complexity are often bundled together, and they are certainly related, as they are both describing nonlinear phenomena. In some ways, however, they are opposite concepts. Chaos Theory, as most of us need to understand it, can be simplified to the idea of sensitivity to initial conditions, commonly represented by the idea that the flapping of the wings of a butterfly in Brazil can bring a hurricane to New York.

2.4 Second Law of Thermodynamics

The second law of thermodynamics states that the total entropy of an isolated system can never decrease over time.

Many of the ideas of complexity are rooted in this physical law, which is often expressed as the idea that you can never unscramble an egg. The sense is that there are many more states of the universe where the system lacks order (more entropy), and thus over time, the system is more likely to go towards increased disorder (entropy), than to an uncommon orderly situation (information). It is in this sense that we can understand complexity and chaos as opposite concepts, for chaos speaks of how disorder can be created even from the slightest deviation in the initial conditions, and complexity deals with how order can be created amidst this disorder, even if only in a local way. It is interesting to think about us, human beings, in light of this dichotomy. As the ultimate representatives of information (the opposite of entropy), we can only exist because we live in an open system, our earth's biosphere, which is fueled by the sun. We consume order and information through our food, and out comes entropy. In other words, just like the Hindu god Shiva (the destroyer), we help the universe go towards entropy faster.

2.5 Emergence

In philosophy, systems theory, science, and art, emergence is the condition of an entity having properties its parts do not have, due to interactions among the parts. Emergence plays a central role in theories of integrative levels and of complex systems.

This is the crux of our matter. We could say that complexity studies emergent phenomena.

2.6 Non-reducibility/Non-reductionist Science/Holism in Science

Systems are approached as coherent wholes whose component parts are best understood in context and in relation to one another and to the whole. This practice is in contrast to a purely analytic tradition (sometimes called reductionism) which aims to gain understanding of systems by dividing them into smaller composing elements and gaining understanding of the system through understanding their elemental properties.

As stated in the previous section, traditional reductionist science attempts to capture all the relevant aspects of aggregate behavior by dissecting it and understanding it as the simple collection of the individual behaviors. For those of us in complexity trying to explain to someone who is not familiar with our holistic approach, the phrase "*the algorithm is the shortest solution*" is a useful way of explaining that the phenomena we study cannot be reduced to simplistic representations. In other words, a dataset reflecting our complex reality cannot be fully described by a linear regression, where variables are accompanied by a positive or negative coefficient that determines the effect of a change of that variable on the system, regardless of the context in which this change would take place.

2.7 Feedback Loops

Feedback occurs when outputs of a system are routed back as inputs as part of a chain of cause-and-effect that forms a circuit or loop. The system can then be said to feed back into itself. The notion of cause-and-effect has to be handled carefully when applied to feedback systems.

In [12], Karl Johan Åström and Richard M. Murray state that: "*Simple causal reasoning about a feedback system is difficult because the first system influences the second and second system influences the first, leading to a circular argument. This makes reasoning based upon cause and effect tricky, and it is necessary to analyze the system as a whole.*"

Feedback loops, in particular positive ones (such as when one action magnifies the next one, just as the action of an investor selling a stock makes the next one more likely to sell as well) are the drivers of much of the emergence we observe in social and other systems. Negative feedback loops, on the other hand, are generally associated with reductionist science, as they make any interesting behavior die out as it percolates through the system.

2.8 Technological Lock-in/Path Dependence/Accidents Matter

Technology lock-in is a form of economic path dependence whereby the market selects a technological standard and because of network effects the market gets locked-in or stuck with that standard even though market participants may be better off with an alternative.

This phenomenon is often a result of positive feedback loops. The more people speak the language that you speak, the more valuable your language is. For this reason, a language that may have originally not been a particularly efficient one for conveying information may become prevalent if enough people speak it at first. This is a concept that comes in various forms and can be conflated with path dependence or the phrase that "accidents matter," in the sense that what may have been originally an accident sets the system in a historical path that does not come back to any type of average or norm. As discussed later, this is a concept that is vastly understudied and deserves further attention and understanding (see the original treatment of this idea in [13]).

2.9 Fat Tails

A fat-tailed distribution is a probability distribution that exhibits a large skewness or kurtosis, relative to that of either a normal distribution or an exponential distribution. In common usage, the term fat tailed and heavy-tailed are synonymous, but different research communities favor one or the other largely for historical reasons. Fat-tailed distributions have been empirically encountered in a variety of areas: physics, earth sciences, economics and political science.

Fat tails refer to the fact that complex phenomena often exhibit more than their fair share of extreme occurrences. This is a product of positive feedback loops, but also of the presence of power laws, where a snowflake can trigger an avalanche, which is related to the concept of the Edge of Chaos. Fait tails and extreme events are to be expected in many nonlinear systems and complex phenomena, but they are yet to be completely understood and further study is needed.

2.10 Power Laws

In statistics, a power law is a functional relationship between two quantities, where a relative change in one quantity results in a proportional relative change in the other quantity, independent of the initial size of those quantities: one quantity varies as a power of another. For instance, considering the area of a square in terms of the length of its side, if the length is doubled, the area is multiplied by a factor of four.

Power Laws are vastly studied within the nonlinear paradigm, but they are arguably not too well understood. One of the most common forms of Power Laws is that of Zipf's Law.

Zipf's law is an empirical law formulated using mathematical statistics that refers to the fact that many types of data studied in the physical and social sciences... For example, Zipf's law states that given some corpus of natural language utterances, the frequency of any word is inversely proportional to its rank in the frequency table. Thus, the most frequent word will occur approximately twice as often as the second most frequent word, three times as often as the third most frequent word, etc.: the rank-frequency distribution is an inverse relation.

The reason why these relationships often appear in nonlinear phenomena is not completely understood, but it is likely a result of the hierarchical way in which many of these systems are naturally arranged. Further study of these phenomena, with appropriate investigation of its origins and not just of its appearances, is warranted in the field.

2.11 Edge of Chaos

The term edge of chaos is used to denote a transition space between order and disorder that is hypothesized to exist within a wide variety of systems. This transition zone between the two regimes is known as the edge of chaos, a region of bounded instability that engenders a constant dynamic interplay between order and disorder. Even though the idea of the edge of chaos is abstract and unintuitive, it has many applications in such fields as ecology, business management, psychology, political science, and other domains of the social science.

This is a concept that was developed in the early years of the complexity revolution, but that has not been properly studied since, and thus represents a potential interesting area of study. The idea is that complex systems such as life and a healthy modern society must live at the edge of chaos—an area where the system is not too rigid, so as to allow innovation (exploration), but also not too lax, so as to allow for the conservation of useful information (exploitation).

2.12 Exploitation Versus Exploration

This is one of the most important ideas riding on the wave of the complexity scientific revolution, and much more discussion will be devoted to it in the next two sections. It is arguably one of the most important lessons that complexity teaches us, and yet a definition cannot be currently found in Wikipedia. For those of us who believe that this tradeoff is incredibly important, and a rich area of research and application in the M&S field, the term itself needs revision, as it seems like there could be a catchier phrase to be used here.

3 Agency in a Nonlinear Setting

The common historically reductionist past in science has increasingly led way to a nascent holistic perspective of nonlinear science that is capable of describing multiple levels of reality. We argue here that novel languages for describing human behavior and social phenomena are necessary. These new ways of communicating must be set within a general theory of collective behavior and structure formation, with a resulting architecture that can be broadly applied. Only in this way will we achieve the necessary consilience across disciplines that is still lacking [14].

Systems in which multiple, heterogeneous, interacting entities adaptively influence the behavior of each other quickly became intractable for traditional mathematical techniques. To continue the advancement of science and technology, most disciplines generally opted for the simplification of interacting agents, commonly describing them as homogeneous, independent and altogether exogenous. As such, the paradigm of Newtonian physics—where atoms interact in predictable patterns to compose aggregate objects—was followed by the classical economic paradigm of Adam Smith [15], in which selfish agents acting independently are guided by an invisible hand to maximize the greater good [16]. Similarly, Darwin's [17] revolutionary theory of evolution describes independent agents that survive based upon their individually defined fitness. These three paradigms delineated the boundaries of most research done for more than two centuries in all physical, biological and social realms—even though they represent branches of the same tree; a tree rooted in a linear description of the world around us.

Standard economic representations have focused on instantaneous individuals or current profit-maximizing firms as the smallest unit or the ultimate irreducible atom of the paradigm, but we propose further attention to the possibility that such units may actually be agglomerates, and thus the products of internal networks that deserve attention. We can think of a company's organization or a strategic military coalition, and no agency is allowed for identifiable players forming the organizations and alliances, nor is emphasis given to the study of blurry abstract borders dividing the participating members. The possibility of understanding some of the internal complexities that give rise to their positions is essentially ignored. One area of potential fruitful research would break apart the threads that make up an agent, whether a person or company, both in an intertemporal and a static sense, in order to recognize the emergent properties that compose it.

To describe a more general theory of social interaction, we need models in which we drop customary assumptions made in traditional disciplines about what is considered a decision-making unit. The relevant agents in the proposed system are intermediate in the sense that they are both influenced by an upper level with its own degree of agency, while at the same time they are determined by relatively independent subcomponents that must be 'subdued' into acceptable behavior. Any observed action is considered to be the result of the interplay of multiple distinguishable actors. The proposed architecture may be used to build large-scale simulations, as well as models that focus on the interaction between levels of interest.

This language must allow us to deconstruct intent, based on the idea that an agent's behavior can be seen as an emergent property of a collection of intertwined aims and constraints. The intent is to establish a way in which we can refer to many different levels of reality with one general methodology. Such a general language is needed given that, after all, the levels of reality have the common thread of being based on a singular reality. We consider a disentangled agent that is formed by multiple and relatively independent components. Part of the resulting agent's task is to present alternatives, or 'fields of action' to its components. Correspondingly, the composed agent is itself constrained by a field of action that the superstructure to which it belongs presents and can thus be ascribed agency and modeled as an agent. To arrive at this, we must redefine what a unit of decision is by unscrambling behavioral influences to the point of not being able to clearly delineate what the individual is, who is part of a group and who is not, or where a realm of influence ends; the boundary between an individual self and its social coordinates must be dissolved. The proposed intermediate agent can be thought of as a person, a family, a social class, a political party, a country at war, a species as a whole, or a simple member of a species trying to survive.

Microeconomics has usually relied on simplistic definitions of what represents an indivisible actor in the two main levels of interaction: Individuals or households for consumption decisions and firms for production ones. Generally speaking, firms have a straightforward behavioral directive: to maximize their profits. Correspondingly, individuals attempt to maximize their utility function. On the other hand, macroeconomic models usually rely on linear aggregation or recursive processes. Economics has thus most commonly modeled behavior through a closed objective function that the agent attempts to maximize, subject to the constraints imposed by the exogenous environment. As the research of Herbert Simon points out (see the interesting take on Simon's marvelous and arguably unfinished research agenda by Hatchuel [18]), however, many economic structures do not organize themselves in a market scheme, but rather optimize their underlying internal relationships with intricate structures of command. Thus, the company is in a constant reorganization process to create a structure that maximizes profits (its objective function or an aggregate measure of 'fitness'), and then turns around to allocate the spoils among its participating members.

Standard economic representations have focused on individuals as the smallest unit or the ultimate irreducible atom of the paradigm, but we propose further attention to the possibility that such units may actually be agglomerates, the products of internal networks that deserve to be considered. An investigation into a hierarchical decomposition of agency reveals that important margins are overlooked when the physical person is the unit of measurement and its corresponding underlying efficiency. The arbitrariness of the use of the static and isolated individual as a unit of measure can be demonstrated by asking ourselves simple questions: Shall we refer to ourselves as the one we are in this very moment? Should we stop caring about our children? If we can think of ourselves in the present and in the future as 'us', then why is our caring for our children any different?

Evolution is therefore not a process in which autarkic individuals compete to be the fittest, but rather a historical multiplex of intertwined levels. In this sense, the phrase "survival of the fittest" takes on a more specific meaning; it does not only mean that the genes of the best-suited individuals in a population will survive, but rather that the populations themselves will become more adaptable, deadlier, more complex, and all together 'more fit' at exploiting environments. To clarify this point and to make a personal contribution to the timeworn tradition of using sports metaphors to defend an idea on evolution, let us conceive of natural selection as a process which seeks to assemble a Super Bowl-conquering football team. What is important to recognize is that the rules of the game are themselves being selected, and it is in this sense that we can talk about more fit individuals, but also about more fit selection processes.

What is special about the organisms that inhabit today's earth? For one, they reproduce in one way or another. There is the possibility that in the past, a highly advanced species decided not to do so, but we would have to hope for paleontologists to let us know of their curiously ephemeral existence. This is one restriction that applies to any existing species. Countless other restrictions apply to each specific species, without which their populations would not survive long. Since these restrictions are external to the individual organism, they may be modeled as upper levels. Natural Selection applies, then, not only to individuals, but just as well to groups, laws of behavior, species, and groups of species. Because there is no visionary designer looking out for the correct implementation and discovery of these upper levels, it is plausible that a group as a whole will behave in ways that do not maximize its potential to keep its recursive process going; however, those that do happen to follow strategies optimized with a longer term horizon are more likely to be observed. This idea bears some resemblance to the popular Murphy's Law, whereby, for example, we can say that you will most likely find the slowest lane in traffic, if only because when you do, you have a lot of time to think about it.

We end this section by referencing the tradeoff between Exploration and Exploitation. The idea is that any complex problem that cannot be solved with conventional optimization methods or numerical techniques must be solved by managing the tradeoff between these two forces: Exploitation (the force of individuality) and Exploration (the force of collectivity). Imagine such a problem as a giant lattice with valleys and peaks. Somewhere in there lies Mt. Everest, the global optimum for this particular problem, but throughout this space lie innumerable hills which represent improvements over the valley next to them, but nonetheless only local optima. The tradeoff between exploration and exploitation means that in order to find the global optimum, the search must find a way to search the valley, recognizing that when there is a positive slope, we must exploit it, go up and find how high it goes. But we must not get immediately stuck in the first hill we find, for we must continue exploring to see if there are no other higher mountains to climb. Such a problem can be something as impossible to solve as to how to best organize society. It can also be the search for the love of your life. The moral of the story is that you should not necessarily marry the first person you kiss, for even if it is a pleasurable experience, this may be an obvious case of too much exploitation. Conversely, you must recognize the impossibility of improving upon finding your perfect mate, for someone who is too much on the side of exploration will continue to move from partner to partner, and never recognize that they have indeed found "the one".

How does the upper level or environment "communicate" to its inhabiting agents where optimality lies? Can it play a game of "cold, warm, hot" with them? Are there threads that communicate where optimality lies that are invisible to some but visible or intuitively attractive to the fittest? The management of this tradeoff is one of the main ways in which we can use M&S methodologies to explore immense fields to find the global optimum. Computer simulations can help us explore ways to build coalitions to tackle climate change, a better drawing of congressional districts in the US, or the way different drugs may interact with one another. The field of M&S has delivered a tool with which we can explore scientific areas that used to be completely in the dark. Let us find the keys where they really are.

4 How Does M&S Incorporate Nonlinearity: What Should We Simulate?

As we have argued, all agents—whether organisms, households or nation states—are incomplete, for they only represent a link in the long recursive process that gave them life, and that forces them to maintain it. To have a meaningful life in the evolutionary sense, sexual animals are forced to find a partner and mate. Many species including all mammals are even further constrained by the fact that the younglings need constant care as they grow up. In a social species, where the optimal scale of environment exploitation requires the coordination of more than one individual, a peer may possess aspects of both a competitor and a potential partner. The optimality of the upper level correspondingly implies subdividing the agents until it finds cohesive subcomponents that are cooperating or competing in nature. This phenomenon appears in treaties such as the Geneva Convention, where countries that expected to have further bloody wars accepted common laws that would benefit the global upper level. In other words, a process of morphogenesis will insure the appearance of structures that better exploit an environment, even if these require constant energy to achieve homeostasis [19].

Organisms thus represent the current stage of long, recursive processes that have searched vast phase spaces to find the combination of complicit levels that sustain the information they entail. Random mutations are generally not the best tool for such processes; rather, controlled ways of adapting to an ever-changing environment will naturally develop. In other words, exchanging a known successful situation for an alternative that will be drawn from an infinite space of unknown possibilities is unfeasible—too chaotic. For example, the emergence of sexuality provides a systematic way for the mixing of populations with information that has proven useful and constantly recreates the organism in order to stay ahead of predators and parasites. It also creates within-species heterogeneity which represents as many experiments as there are individuals, always looking for the best forms to exploit a given situation. We could think of an enhanced sexuality that provides other improvements to the population, such as superior genetic material that is phenotypic and recognizable. The benefits of sexuality can come from different levels; at a localized and instinctive level, in which mates with compatible genetic structures might produce smells that create attraction, or at the less-localized level, with non-genetic social norms that lead to correct population control.

The focus of this essay is humans, but the general idea of this discussion should be applicable to any species. The important point here is that humans are alive because they are fit to be so in innumerable ways. In line with the theme of this work, a human's utility function is not created in a vacuum but is instead the product of a culture and millions of years of natural evolution. One of the many reasons why humans have been so successful is the fact that we learned how to cooperate with each other. In this perception, the space of possibilities before the advent of Humans included many different possibilities for an emerging species to exploit.

If we optimize that space in search of what kind of species would be best fit for such an environment, what we would find is that a cooperative one would be better equipped than its non-cooperative counterpart. In this sense, the optimized phase space of the environment represents a series of demands for exploitation, while the existing species represent a series of supplies. When the two meet, a recursive and self-sustaining entity is produced. In this sense, we can think of evolutionary pressure as a situation in which one group of individuals is in position to mutate and fill up a niche that the environment presents. Social structures must then be analyzed in the same fashion we would think about the limb of an organism: the way in which evolution acts upon a social network must be studied, just as an ethologist studies animal behavior. Evolutionary pressures shall act upon individuals and subcomponents of an individual, but just as well upon the structures these individuals form.

The Law of Requisite Variety, sometimes known as Ashby's Law, implies that the variety in the control system must be equal to or larger than the variety of the perturbations in order to achieve control. To achieve homeostasis, the agent must therefore be at least as complex as its environment. Similarly, when we think of our models and the level of description at which they should be built, we must first understand what it is that we are trying to model. A good model should capture all of the ways in which the system and its component parts or agents can react. In other words, a model of a complex adaptive system must capture all the adaptability present in the phenomena at hand. A simple Multi-Agent System in which the agents interact may exhibit emergent behavior, and thus capture many of the relevant aspects of reality, but it will likely nonetheless fail to capture its multidimensional nature. Our practice should move towards more multi-leveled simulations that can capture these essential aspects of complex systems.

5 A Call to Action

We end this article with a call to action directed to all of us working in the Modeling and Simulation arena, both in theoretical fields as well as in applications of the methodologies. More than proposing recipes for future work, this concluding section leaves a series of questions open for the reader to ponder upon. The hope is that as the years go by, more and more of these questions will have found an appropriate answer.

- What are we doing to implement some of our findings in the real world?
- Where else can we search for possible applications of the M&S methodologies in the real world?
- How can our simulations benefit humanity?
- How do we know that a simulation is a good representation of reality? In other words, what are we doing as a field to assure the epistemological virtues of our work?
- How do we separate a good from a bad implementation of a model? How are these applications vetted?
- How are our models informing the primary data that needs to be collected from the world? Are we finding appropriate ways to incorporate all of the different types of available data?
- How are we summarizing in our field the types of methodologies that are appropriate in a successful simulation? Are there guidelines that allow researchers to understand the areas of applicability for each methodological procedure?
- Are our methodologies in the field of M&S effective and complete? Are we still in a process of cleansing our procedures?
- Can our work be replicated by others? Can we corroborate our findings?
- How are the scientific disciplines incorporating the concepts of complexity science?
- Are traditional disciplines accepting the use of M&S as a valid tool of scientific inquiry?
- Can we find ways to make scientific disciplines more consilient?
- Based on the idea of emergence, complexity is a natural conduit for interdisciplinarity. Can complexity and systems thinking help us develop models in which we incorporate the perspectives of two or more disciplines or paradigms in one common description?
- If a multi-level model can describe the reigning paradigm in more than one discipline, how can it help the two involved disciplines to better inform each other?
- Can our complicated models express the more simple or lineal models that are currently used in traditionally linear disciplines? Can we show that the simpler models of the past are nested in our complex ones?
- Are we educating our students on the lessons of complexity and the benefits of simulations?

We posit that there are vast amounts of work left in our field. Let us get to work, for the rewards will be great. The revolution that complexity and simulation are creating promises to bring fundamental changes to the way that science is done. For this revolution to take hold, this chapter represents a call to action for all of those working in the field to find the missing pieces of the puzzle. Only when we can be comfortable in the soundness of our methodologies and the clarity of our concepts will we be fully ready to express the benefits of our approach to others who may not understand its benefits at first glance.

References

1. Tolk, A.: Limitations and usefulness of computer simulations for complex adaptive systems research. In: Sokolowski, J., Durak, U., Mustafee, N., Tolk, A. (eds.) Summer of Simulation 50 Years of Seminal Computing Research. Springer International Publishing AG, Cham, Switzerland (2019)
2. Bar-Yam, Y.: A mathematical theory of strong emergence using multiscale variety. Complexity **9**, 15–24 (2004). https://doi.org/10.1002/cplx.20029
3. Abbott, R.: Emergence explained: abstractions: getting epiphenomena to do real work. Complexity **12**, 13–26 (2006). https://doi.org/10.1002/cplx.20146
4. Pessa, E.: What is emergence? In: Minati, G., Pessa, E. (eds.) Emergence in Complex, Cognitive, Social, and Biological Systems, pp. 379–382. Springer, Boston, MA (2002)
5. Crutchfield, J.: The calculi of emergence: computation, dynamics and induction. Physica D **75**, 11–54 (1994). https://doi.org/10.1016/0167-2789(94)90273-9
6. Taylor, P., Jonker, L.: Evolutionary stable strategies and game dynamics. Math. Biosci. **40**, 145–156 (1978). https://doi.org/10.1016/0025-5564(78)90077-9
7. Goldstein, S., Young, C.: Evolutionary stable strategy of handedness in major league baseball. J. Comp. Psychol. **110**, 164–169 (1996). https://doi.org/10.1037/0735-7036.110.2.164
8. Lindgren, K.: Evolutionary dynamics in game-theoretic. In: The Economy as an Evolving Complex System II, p. 337 (2018)
9. Axelrod, R.: More effective choice in the Prisoner's dilemma. J. Conflict Resolut. **24**, 379–403 (1980). https://doi.org/10.1177/002200278002400301
10. Gilboa, I., Samet, D.: Bounded versus unbounded rationality: the tyranny of the weak. Games Econ. Behav. **1**, 213–221 (1989). https://doi.org/10.1016/0899-8256(89)90009-2
11. Allen, T., Starr, T.: Hierarchy: Perspectives for Ecological Complexity. University of Chicago Press, Chicago (1982)
12. Aström, K., Murray, R.: What is feedback? In: Feedback Systems, an Introduction for Scientists and Engineers. Princeton University Press (2010)
13. Arthur, W.: Competing technologies, increasing returns, and lock-in by historical events. Econ. J. **99**, 116–131 (1989). https://doi.org/10.2307/2234208
14. Suarez, E., Castañón-Puga, M.: Distributed agency. Int. J. Agent Technol. Syst. **5**, 32–52 (2013). https://doi.org/10.4018/jats.2013010103
15. Smith, A.: The Wealth of Nations. Modern Library, New York (1776)
16. Beinhocker, E.: Evolution, complexity, and the radical remaking of economics. In: The Origin of Wealth. Harvard Business School Press, Boston (2006)
17. Darwin, C.: The Origin of Species by Means of Natural Selection: Or, the Preservation of Favoured Races in the Struggle for Life and the Descent of Man and Selection in Relation to Sex. Modern Library (1872)
18. Hatchuel, A.: Towards design theory and expandable rationality: the unfinished program of Herbert Simon. J. Manag. Gov. **5**, 260–273 (2001)
19. Maturana, H., Varela, F.: Autopoiesis and Cognition, the Realization of the Living. Springer, Netherlands, Dordrecht (1980)

Printed by Printforce, the Netherlands